INNOVATE WITH GLOBAL INFLUENCE

Tales of High-Tech Intrapreneurs

By

Steve Todd

BookLocker.com, Inc.
2010

To my co-workers past and present

Acknowledgements

Special thanks to Chuck Veit and Nancy Rosenbaum

for their help with graphics and editing, and to the EMC Volante team

for their assistance with global coordination.

Table of Contents

List of Figures

FOREWORD

Vijay Govindarajan

As I stood on the stage of the Nokia Theater in New York City, I looked out at a crowd reeling from the effects of the ongoing global economic crisis. It was May 2009, and the attendees had come to the World Innovation Forum hoping to gather actionable insight on innovation in the new world order. Changing times require radically new strategies.

To my left, in the balcony positioned twenty feet above the rest of the crowd, sat Steve Todd and the rest of the bloggers covering the conference. Steve was not only sharing his impressions of the conference with the rest of the world; he was also formulating his own action items to bring back to his multi-national corporation, EMC®.

In the months that followed, Steve and I began an exciting collaboration on one of the most critical business themes of our time: reverse innovation.

As professor of international business at Dartmouth College, I am witnessing a great shift in global markets. The income gap between the United States and developing countries is so large that products designed for American consumers simply cannot be adapted to, for example, Chinese or Indian mass markets. These locations, however, are expected to experience the largest expansion of any markets in the coming decade. Large, multi-national corporations headquartered in wealthier locales simply cannot be globally competitive unless innovation occurs natively – in the developing countries.

The essence of reverse innovation is that products developed locally are distributed globally. It is a formidable organizational challenge for incumbent multi-nationals in wealthy countries to operate according to this model. Entrenched intrapreneurs at corporate

1

headquarters must share (or cede) creative control to their satellite counterparts.

This phenomenon of far-flung innovation is especially true in the high-tech industry. The amount of digital information being generated worldwide is staggering, especially in developing countries. Corporations like EMC specialize in this market, and already employ R&D teams in many of these countries. These locations were established as part of a sound plan of globalization; EMC expanded its global presence while minimizing costs.

In my role as Chief Innovation Consultant at General Electric, I consulted on the development of a global innovation framework that unlocked the innovative potential of employees worldwide. This new framework received strong support and implementation at the highest executive levels.

What I find inspiring about EMC is that the cultural shift to reverse innovation is being largely driven by the employees in the trenches! This book is filled with stories of global intrapreneurs who consistently reach across geographical boundaries to deliver breakthrough product innovation.

Steve's storytelling is significant in that he is also in the trenches, right alongside his co-workers. Whether they be in Russia, Ireland, Israel, India, China, or Egypt, Steve and his co-workers recognize the global opportunity in front of them. No stranger to innovation himself (he is named on 160+ patent applications!), Steve describes a road to reverse innovation that relies on the mentoring and transfer of intrapreneurial skills to global sites.

Innovate With Global Influence is more than a collection of inspiring stories of innovation. It is both an employee handbook and a corporate framework for reverse innovation.

Vijay Govindarajan
Professor of International Business
Dartmouth College

SECTION I: GLOBAL PRESENCE

1. Introduction

At 6 p.m. in Cambridge, Massachusetts, a software engineer finishes his day. He starts a download of a video clip before he heads for home.

Bits from the video clip begin to travel a rather unique journey. They are copied from his laptop onto a network. They join the Internet via a network switch, and ultimately reach the intended destination on the inside of a storage system housed in the back room of a test lab in Hopkinton. His information is safely stored in no time at all.

Then it gets interesting.

The storage system begins to intelligently analyze the information. It determines the bits represent a video file. It recognizes the creator of the video file. And it uses this information to determine where else in the world this information might go.

Video data written by this particular engineer are dynamically routed to two additional locations: Santa Clara, California, and Shanghai, China. Software engineers in those facilities are still at work (or about to show up). The storage system automatically distributes the information to both locations. Any eventual edits made in these locales will automatically travel back to Cambridge.

The information, in this case, follows the sun. Other types of information, such as spreadsheets and business presentations, might get sent to London, or to Singapore.

This information storage system is unlike any of its predecessors. Engineers who build these sophisticated products play a starring role in the new age of digital expansion. More than ever before, they need to find solutions to match the explosive pace of information growth today and tomorrow.

Customers of information technology products face challenges as they strive to cope with growing quantities of data. Continuous acquisition of new storage devices is expensive, and more than just the purchase price must be considered. System administrators must be paid to manage all that information. Government agencies' information audits can trigger huge fines if fault is proven. The electricity required to power information storage devices results in skyrocketing utility bills.

These problems are experienced all over the world. Engineers can no longer solve these problems solely within the confines of corporate headquarters in the United States; the growing markets are in developing countries. Large corporations that deliver high-tech products need innovative engineers in locations around the globe. These engineers can interact directly with their local customers and build a better product to meet a particular locale's needs.

Many people believe entrepreneurs from native start-ups will have a better chance of penetrating these markets with new high-tech gadgets than will foreigners. That may be true. Do multi-national corporations stand a chance? They do. If they can build a network of globally distributed innovators, the resources they bring to bear can be a powerful advantage.

Corporations with a global presence need to take advantage of intrapreneurs in developing countries. Intrapreneurs are employees at large corporations who have the unique ability to consistently deliver their own ideas. Like their counterparts – the entrepreneurs – these passionate creators can motivate their teammates to innovate and build.

Intrapreneurs in developing countries are motivated by something their U.S.-based counterparts do not possess: desperation. Customers in developing countries have strong needs for inexpensive technology that improves their lives. This desperation, either experienced or directly encountered by intrapreneurs, can fuel innovation in ways not possible in wealthier locations. Low-cost products developed to address local needs may possess attractive price points, adding to their import appeal in countries such as the United States.

As a new wave of university students begins to explore job opportunities, they may consider taking college courses in entrepreneurship to increase their odds of landing an innovative job. If they are lucky, they may be exposed to the concept of an intrapreneur in a "corporate entrepreneurship" course.

Introduction

What many of them are not being taught, however, is that an innovative career may best be pursued at a large, global corporation. The changing economic landscape favors multi-national corporations that already have a large R&D footprint in the world's developing countries.

In this sense, a young intrapreneur has a distinct advantage over a young entrepreneur: a global experience.

This option goes against the grain of conventional thinking. Aren't start-ups filled with energy and passion, while corporations are filled with obstacles and red tape?

It depends.

This book is filled with the stories of the global intrapreneurs who work for EMC Corporation. Each story describes a young innovator who has taken the risk of proposing an idea and then stood behind it and delivered.

The stories, when taken as a whole, serve as a template for intrapreneurial behavior. They are ensconced within a corporation that has implemented a new, global management framework that, by necessity, decentralizes innovation. Idea generation is a level playing field and the opportunity to deliver a global idea is there for the taking by any employee who practices intrapreneurial qualities and characteristics.

When the sun sets on the United States, it starts to rise in China.

That's when the intrapreneurs begin showing up to work.

2. The World Goes to Work

When the sun rises on mainland China, Hang Guo begins his short commute to Zhongguancun, a technology district located in northwestern Beijing. Zhongguancun has often been referred to as "the Silicon Valley in China." As Hang rides the bus to his office at the Tsinghua Science Park, he passes by two of China's most revered universities: Peking University and Tsinghua University. Figure 1 depicts the high-tech neighborhood through which Hang travels each working day.

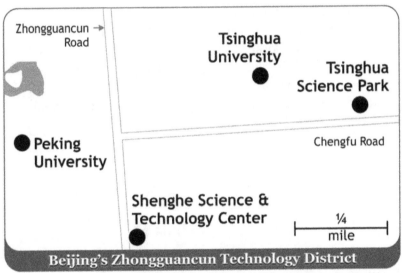

Figure 1. Co-located high-tech centers and academia in Beijing.

The co-location of industry and academia is integral to Beijing's reputation for high-tech innovation. These universities have long been heralded as two of the top learning centers in all of China. Application for admission is strictly limited to the top scorers from China's National Entrance Examination, and only the most elite are accepted. Graduates from these institutions go on to play important roles in nearly every area of China, producing an impressive array of achievements including publication in scholarly journals and presentations at international conferences. Both universities have excellent computer science departments.

Hang is a recent graduate of Tsinghua University. His wife graduated from Peking University. The two are frequent visitors to the Peking University library. With its 51,000 square meters, more than 5,000 seats, and in excess of 6,500,000 items, the library is the largest of any university in all of Asia.

Hang works in one of the most innovative, high-tech environments in the world. Learning and research are revered in Beijing. When he chose to work at EMC, he did not settle for a simple job maintaining software written by someone else.

He wanted to build something new.

At sunrise and approximately 600 miles to the south of Beijing, Hang's corporate co-worker, Viki Zhang, is making her way to her office in Shanghai. EMC has multiple worldwide centers of excellence (COEs). Each COE has a strong workforce of R&D experts. Figure 2 depicts the distance between the Beijing COE and the Shanghai COE.

Figure 2. EMC researches and develops high-tech products in both Beijing and Shanghai.

The World Goes to Work

Shanghai is not only one of the biggest cities in China; it is also one of the largest cities (by population) in the world. It is a massive center of commerce and finance, and serves as a favored locale for thousands of domestic and international companies. Young people flock to the city to find jobs in the booming economy. One of these people is technologist Viki Zhang.

Viki is an innovator and a problem solver. She enjoys collaborating on the building of next-generation high-tech products and solutions for her country and her world. Before she starts each workday, however, she has a very practical problem to solve: how to get to work.

Shanghai's public transportation options are numerous. Whether the vehicle of choice is a bus, train, boat, or car, the city is a massive criss-crossing of railroads, bridges, channels, highways, and subways. Many companies offer free shuttle bus services to facilitate their employees' commute to work. When she first started working, she lived on the outskirts of the metropolis. As her responsibilities for interaction with her global account team grew, she moved deeper into the city and closer to her job.

And she bought a small, red bicycle.

The destination for her bike ride is the Wu Jiao Chang commercial area of Shanghai's Yang Pu District, as shown in Figure 3.

Figure 3. Shanghai's Yang Pu District is ideally situated for bicycle commuters.

Viki's commute takes her past Jiangwan Stadium (formerly known as Shanghai Stadium). Completely renovated in 2007, the stadium has hosted such major events as the Chinese X Games and the Special Olympics. On most mornings, she exchanges pleasantries with the neighborhood's many fitness enthusiasts who engage in morning exercises. The hustle and bustle of this thriving community energizes and prepares her for innovation as she arrives at her office park.

Over 3,000 miles away, in Bangalore, India, Arun Narayanaswamy begins his 10-minute drive to work. Everywhere he looks, he sees construction to facilitate travel and enterprise. There are flyover bridges being built, as well as new streets and new intersections. Commercial buildings are being erected, along with private residences rising up to house the thousands of workers moving into the city to capitalize on the booming economy and job market.

His office is located in a brand new facility on Outer Ring Road, near Marathahalli, a bustling neighborhood with an incredible number of high-tech corporations, including EMC, IBM, Oracle, Cisco, HP, and Google. The EMC facility, located in the Bagmane World

Technology Center, was opened in September 2009 as part of a $1.5 billion investment over a five-year period. In spite of the enormous number of new technology companies and travelers to the region, Arun still stops his car for cattle crossing the highway!

Bangalore's residents and visitors enjoy the city's popular parks and tourist destinations. Cubbon Park, located west of Marathahalli in Bangalore center, is a 300-acre public park that boasts over 6,000 plants and trees, including indigenous and exotic botanical species. The park can be approached via the magnificent Karnataka High Court building, which now serves as the seat of legislature for the area.

Local employees tend to arrive at work via company-provided bus transportation, but motorbikes are an increasingly popular commuting mechanism. Many employees also travel back and forth between the Bagmane World Technology Center and the India Institute of Management. The institute, popularly known as IIM, ranks among the top business schools in the world and attracts some of the brightest talent in the country. Employees take courses to enhance their knowledge of some of the latest and best management practices in the world. Figure 4 depicts the location of Arun's office and the locations of local parks and education.

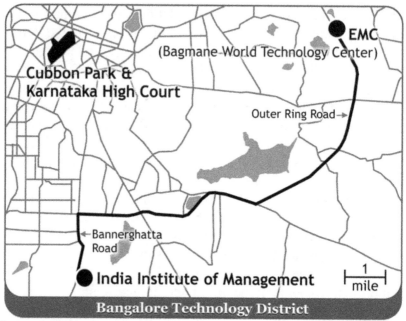

Figure 4. Bangalore's growing community features technology and academia.

Just over 3,000 miles to the west of Bangalore, Assaf Natanzon begins the commute to his office in Ramat Gan, Israel. Ramat Gan is located just east of Tel Aviv. Distance from the office determines how employees commute to Ramat Gan. Assaf travels by foot or by car; many others commute by bike or by train. In Israel, the work week begins on Sunday and extends to Thursday.

Assaf's office is located in the diamond exchange district in the northwestern section of Ramat Gan. The area contains numerous skyscrapers, including the Moshe Aviv tower, the tallest building in Israel. Assaf works in one of multiple EMC offices that make up the Israel COE. These offices are spread out across a 15-mile radius.

Israel has a unique ability to attract venture capital. The Tel Aviv area ranks second in the world (Silicon Valley ranks first) for highest number of startups. *New York Times* columnist David Brooks noted, "With more high-tech startups per capita than any other nation on earth, by far, corporate investors and venture capital firms pour an estimated $1.5 billion into new ventures there each year. Israel, with 7 million people, attracts as much venture capital as France and Germany combined."

Employees travel frequently between the three offices within the country to promote cross-pollination of technologies and experiences. Israel's advanced academic institutions also provide foundations for the thriving high-tech industry. Four of the top 30 computer science universities in the world are located in Israel. Figure 5 depicts the locations of one of Israel's COEs and its proximity to these academic institutions.

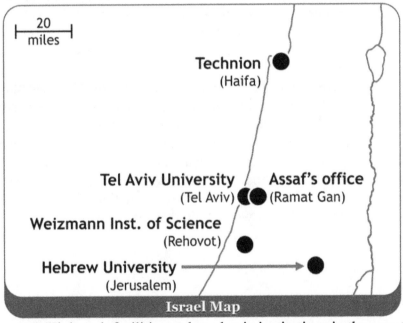

Figure 5. High-tech facilities and academic institutions in the country of Israel.

Just over 1,900 miles to the north of Ramat Gan, Denis Kiryaev rides the subway into St. Petersburg, Russia. The subway itself is one of the most attractive in the world, with many of the stations adorned with exquisite artwork and extensive decorations. The city is located adjacent to the Gulf of Finland; water accounts for 10 percent of the city's area. St. Petersburg is sometimes called the "Venice of the North" due to its many rivers, canals, and waterways. The more than 300 bridges across these various waterways can make commuting to work in St. Petersburg a daunting task. Not all bridges accommodate cars and alternative paths are limited. Traffic jams inevitably occur and add to the subway's appeal with working commuters.

Subways from various directions run deep under the city's landmarks. The historic center of St. Petersburg is the Palace Square, home to Catherine the Great's Winter Palace. Located within the Winter Palace is the State Hermitage, an enormous museum of art, history, and culture that contains artifacts from around the world. The Hermitage is adjacent to the main street, Nevsky Prospekt, a long, wide thoroughfare that boasts a large number of shops, theaters, and restaurants.

Just across the Neva River lie many of St. Petersburg's universities, including two prominent technical schools: St. Petersburg State University, and the State University of Information Technologies.

Commuters heading for the EMC COE in St. Petersburg disembark at the Vasileostrovskaya subway station onto Sredniy Prospekt and walk the rest of their way to work in this bustling business district. Figure 6 depicts the landmarks of St. Petersburg passed by Denis and his co-workers on their way to work.

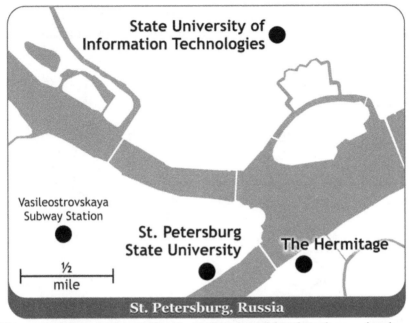

Figure 6. High-tech, academia, and cultural landmarks are in close proximity in St. Petersburg.

Over 1,500 miles to the west of St. Petersburg, Shane Cowman arrives to work in Cork, Ireland. Cork is the second largest city (by geography) with the third highest population in Ireland. No matter which direction he and his co-workers travel from home to work, the commute is picturesque.

Just to the south of Cork are the quaint, narrow streets of Kinsale, the Fransiscan Abbey at Timoleague, and the beaches of Clonakilty (some of the best beaches in Ireland).

Commuters from the west pass through Killarney, renowned for its national park's rolling hills, streams, lakes, and waterfalls. The drive from Killarney is green as far as the eye can see, and roundabouts dot the driver's path to Cork.

The city itself contains two of Ireland's top universities: University College Cork, and the Cork Institute of Technology. Between the high-tech opportunities in the city and the beauty of

Ireland's countryside, it is no wonder the EMC Cork COE has consistently been voted one of the top places to work in Ireland. Figure 7 depicts Cork's location.

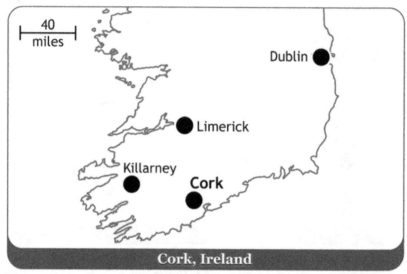

Figure 7. Cork, Ireland, is home to EMC and spectacular natural beauty.

Although their commutes are all different, when these intrapreneurs arrive at work, they share a common objective: innovation on a global scale.

Perhaps our various commutes provide us motivation to not only succeed, but also excel once we reach our place of work. The effort one takes to connect, either via digital network or face to face in the office, provides a springboard to culture, teamwork, and the inherent desire to succeed and surpass our own wildest expectations.

Daniel Pink, in his book *Drive: The Surprising Truth About What Motivates Us*, calls these individuals Type-I personalities. They are not motivated solely by extrinsic carrots (rewards) and sticks (penalties). Their reward is the activity of the work itself. The opportunity to generate (and work on) their own ideas is so valuable to them that they decide to often exceed customers' desired solutions.

It's a good thing that they do, because employee innovation has become a global imperative.

3. The Global Imperative

In May 2009, in the midst of a global economic crisis, a distinguished set of speakers took the stage at the Nokia Theater in New York City. They spoke of innovation as part of the World Innovation Forum. Speaker after speaker stressed the continued importance of innovation during lean economic times. A not-so-subtle theme emerged again and again: The United States is no longer the fastest growing market for high-tech innovation.

In October 2009, in the context of a "jobless recovery" from the economic crisis, a different set of speakers took the stage at Radio City Music Hall in New York City. They were business leaders lecturing at the World Business Forum. A not-so-subtle variant of the same theme was played out: The United States was no longer the world's banker. Much of the money to drive innovation and technology was now held overseas.

Countless high-tech advancements of the last several decades have arisen from the innovative efforts of entrepreneurs in the United States. Startups sprang up on both coasts of the country and tapped into plentiful quantities of venture capital to propel ideas into globally distributed products.

Many of the greatest high-tech inventions were built on the backs of Baby Boomers working on applications, networks, databases, and information storage systems. The combination of these technologies fueled both the growth of the Internet and the dot-com boom.

Ten years later, innovation experts showed up in New York City and asked a fundamental question about high-tech innovation in the United States: "Is it over?"

For startup companies looking to address global markets, it may very well be. A global footprint may be required in the new world order.

What about large corporations? Can they still drive innovation that addresses global markets?

The answer is yes, but only for those corporations that fundamentally alter their internal processes and begin to effectively innovate with global influence. Innovative behavior must be distributed and cultivated throughout their organizations.

In other words, they must create and nurture a workforce of global intrapreneurs, employees who are enthusiastic about generating breakthrough ideas and delivering them to worldwide markets within the context of a large corporation.

Two experts with particular insight into the importance of global innovation spoke at the 2009 World Innovation Forum. Vijay Govindarajan is a professor of international business at Tuck Business School at Dartmouth University. C.K. Prahalad, in one of his last appearances before his untimely death in 2010, was a professor at the Ross School of Business at the University of Michigan.

Both of them cast their messages in the context of global markets. Let's look at the warnings and messages raised by each of these speakers and determine if a corporate network of distributed intrapreneurs has the potential to address each need.

Vijay Govindarajan

Vijay is the author and evangelist of the 3-box strategic approach to corporate innovation. This approach is depicted in Figure 8.

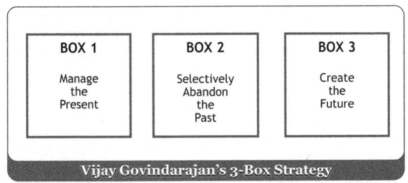

Figure 8. Corporations drive innovation by investing resources in each of three boxes.

Three-box innovation strategy dictates that the majority of corporate resources should be placed in Box 1: Manage the Present. This box represents the continued development of existing products to yield most of a corporation's revenue. Employees supporting this box focus on existing customers and processes, and they continue to leverage their existing competencies. In essence, this box "funds" the development of innovation within a corporation. Some companies fall into the trap of spending close to 100 percent of their resources in this box.

Vijay advises corporations to allocate portions of their resources to Box 2 and Box 3 as well as tried-and-true Box 1. Box 2 selectively abandons the past by "forgetting" most of what is known about the products built in Box 1, including why they were built and whom they were built to satisfy. This break from tradition enables an innovator to take existing products into completely different markets.

Box 3 is a more radical approach to innovation. It completely ignores current processes and products and prominently targets the future.

In the past, Vijay advised corporations to track their investments in these three boxes. For example, a corporation might employ an 85-10-5 approach, or a 75-15-10 approach.

Given the changing global dynamic, Vijay warns that the 3-box approach becomes more difficult to manage. It does not work when

Box 2 and Box 3 activities are based only in the United States. Unfortunately, many companies are in this unenviable position because they have yet to recognize the global phenomenon that Vijay refers to as *reverse innovation.*

Reverse innovation occurs when new ideas and products are generated in developing markets and exported back to wealthier countries. This can occur in large corporations that have progressed through the five phases of reverse innovation, as depicted in Figure 9.

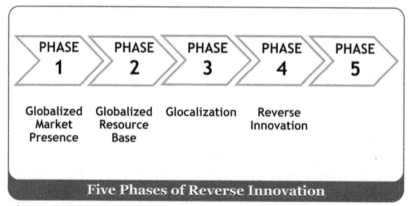

Figure 9. Reverse innovation optimizes intrapreneurial talent in developing countries.

The phases of reverse innovation are as follows:

1. Products are developed in the United States and "dropped" into the world.
2. Global COEs are created that leverage low costs of labor.
3. COEs "glocalize" U.S. products for local markets.
4. COEs begin to invent their own business processes and innovate for local markets.
5. COE innovation is exported globally.

Thinking representative of Box 2 and Box 3 – which includes only wealthier countries – cannot possibly meet the needs of growing markets in developing countries. Those corporations that wish to participate in the new economic reality must, therefore, take advantage of their own intrapreneurs in foreign countries.

Corporations need to choose how to globally distribute the resources that support Box 2 and Box 3. Intrapreneurs, on the other hand, need to apply these boxes to themselves and the local business units of which they are a part. This topic will be explored further in the next chapter.

C.K. Prahalad

In C.K. Prahalad's book, *The Fortune at the Bottom of the Pyramid*, the author discussed the economic opportunities available in developing countries, especially for those countries that are experiencing poverty. He often referred to the poor as "value-conscious consumers." Their strong desire to improve their lot can result in a tremendous drive when they are engaged in the innovation process. CK referred to the engagement of these consumers in the innovative process as the "process of co-creation."

Prahalad presented a framework for poverty eradication. Interestingly enough, this framework can also be applied as a strategy for global profitability for any corporation willing to adopt a new approach to innovation:

> What is needed is a better approach to help the poor, an approach that involves partnering with them to innovate and achieve sustainable win-win scenarios where the poor are actively engaged and, at the same time, the companies providing products and services to them are profitable. (Prahalad, 2010, pp. 27-28)

This philosophy is especially true in the high-tech field. Consider the built-out, high-tech infrastructure currently found in the United States versus that found in developing countries. In the United States, millions of homes have hard-wired network connections and sophisticated power grids. A desperate *need* for innovation in high-tech products is not found in the United States because the country already has an advanced state of infrastructure.

Prahalad explained that technology is readily consumed in developing countries that currently have little to no high-tech infrastructure. In India, the use of cheap wireless devices spread when it was shown that Indian farmers could check prices at local auction houses (as well as the Chicago Stock Exchange!), allowing them to make decisions about how much to sell and when. Inexpensive forms of solar technology are being used to heat water in remote areas of Asian countries where electricity is scarce or prohibitively expensive.

In both India and China, high-tech products are being widely adopted when they meet customer needs at affordable prices.

One of the key elements of Prahalad's theory is that these new inventions must be co-created with the audience for which they are intended. Corporations must work closely with consumers to ensure that new products truly meet their needs.

The market opportunities available in developing countries cannot be addressed by corporations if all of their intrapreneurs are deeply entrenched within their own cultures in a single geographic location. What is required instead is a local intrapreneur who can take the initiative to engage with these customers directly and participate in innovation by co-creation.

For innovation by co-creation to happen, corporations need to identify, train, and empower a set of global intrapreneurs who support each others' efforts. When employees have developed the skills and been given the autonomy to innovate in their own locale, they can engage local consumers and enter into the process of co-creation.

Engagement with local consumers and customers fosters ideas, and these ideas will lead to product proposals. In this key stage of reverse innovation, it is imperative that local intrapreneurs interact with their global counterparts. Why? Because intrapreneurs need to collaborate with each other on the most important high-tech issue on the planet: global sustainability.

All corporations that build new gadgets, whether they are small mobile devices or enormous disk arrays, are increasingly being held accountable for the environmental impact of building, running, and recycling their products. A dizzying variety of regulations is being introduced to address matters of how products are built, how they are run, and what to do with them when they are retired.

The responsibilities of individual engineers to build globally compliant and eco-friendly products have never been greater.

In the book *Citizen Engineer: A Handbook for Socially Responsible Engineering*, authors Greg Papadopoulus and David Douglas call the challenges of building sustainable products "incomprehensible." From poverty to global warming, famine, and slave labor, it is impossible for a localized engineer to build truly sustainable products in one locale without connecting with other engineers throughout the world. As they wrote in a March 5, 2009, article published online by *Financial Times*,

> There is a pressing need for engineers to become more proactive with society – to engage, to communicate, and to lead. We can no longer be content that our laboratory work will be presented and translated to the public as originally envisioned.
>
> The time has come for us to become more socially responsible engineers who not only solve problems but who can educate and argue for their ideals. It's the dawning of a new era, the age of the 'citizen engineer.' (Papadopoulus & Douglas, para. 10-11)

Engineers with a global outlook have a dizzying array of new, global tasks:

- Analyze every piece of hardware in their product, know how much energy was required to build it, where it was built, which nation built it (and the nation's politics), and if the manufacturer will take it back when it's retired from service.
- Analyze the carbon footprint, water usage (e.g., cooling), and overall environmental impact of running the product for

an extended period of time. Know the governmental regulations related to the product across all continents.

- Plan for the end-of-life return of the product and prove that the product was returned and effectively recycled.
- Understand the shortages of materials and the eventual depletion of natural resources that are used to build products
- Enable software reporting tools that allow for measurable compliance with local regulations.

If every engineer in an organization must learn and be trained in sustainable product design for every country in which a product will be manufactured and sold, who can be relied on to gather information from global locales? One answer is to build a network of global intrapreneurs, each responsible for information for his or her native locale.

Global intrapreneurs are highly respected members of the engineering community that have built strong local ties with customers and their needs. They have gained the respect of their organizations because they know how to cut through red tape to deliver innovation. In the context of global sustainability, they have taken the initiative to find and reach out to other global intrapreneurs. A strong, global, intrapreneurial network can be leveraged to build products for the global market.

Building this network becomes a matter of identifying and nurturing intrapreneurs and then connecting them globally. Some intrapreneurs have an innate, natural ability to deliver their ideas. Others can be trained in intrapreneurial behavior.

Either way, a corporation needs to know the seven habits of highly successful intrapreneurs.

4. Seven Habits of an Intrapreneur

Innovation is a team sport. High-tech ideas become products only through collaboration.

The team performs at its best when its key players have a full command and grasp of the traits of an intrapreneur. Global corporations need intrapreneurs who possess a unique set of qualities that permit them to successfully navigate their ideas through the ubiquitous obstacles of large companies.

Each global intrapreneur whose story is included in this book has generated and delivered unique ideas. An individual's ability to consistently and repeatedly deliver innovation in a corporate setting is rare. If, however, the pattern of innovation exhibited by these intrapreneurs could be captured and transferred to the corporation at large, the following benefits would be achieved:

- The corporation would have a portfolio of global innovation. Like a venture capitalist, the executive team could choose to invest in the most promising intellectual property.
- Employee engagement and retention would improve. When people are passionate about their own ideas (and are allowed to work on them), the mood of the workforce rises.
- The corporate brand of "a great place to work" would naturally self-promote. This concept is especially true when intrapreneurs participate in external social media. When the innovative culture of a corporation is communicated in human terms, it can help to attract and retain top talent.

It is worthwhile, therefore, to summarize the pattern of innovation practiced by individual global intrapreneurs. This pattern can be used to train global employees, and it can also be used to establish the foundation of an intrapreneurial mentoring program.

The training of individual intrapreneurs must be accompanied by the training of managers (low-level and executive). A framework for global coordination of intrapreneurs will be covered in a later chapter. This chapter focuses on the desired traits that make a great intrapreneur.

Habit #1: Productivity

Figure 10 depicts a great intrapreneur's core competency: productivity. Each intrapreneur must be known as a go-to-person. Great intrapreneurs deliver high-quality results and are ahead of schedule, even when given aggressive work assignments. They are experts in their "sphere of expertise."

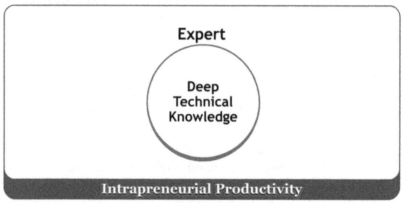

Figure 10. Intrapreneurs use extraordinary productivity to develop deep technical knowledge in a given area.

While an intrapreneur is necessarily a highly productive employee, not every highly productive employee is an intrapreneur. Some productive employees like to work hard, establish a deep understanding of a technology, and then *stay there*. They are more than content to perpetually tweak and enhance their creations (or perhaps re-architect the same technology into a different form). This type of employee is indispensable for sustaining revenue generation from successful products. Without them, corporations could not fund newer, more innovative proposals. These employees reside comfortably and happily in Vijay Govindarajan's Box 1.

Corporations typically recognize and reward the efficiency and obedience of productive employees.

Intrapreneurs, on the other hand, are productive but not satisfied to simply stay in their sphere. They remain in place long enough to deliver on their commitments, but they are intrinsically driven to create something new. They put into practice the second habit of intrapreneurs: initiative.

Habit #2: Initiative

Without question, initiative is the key attribute that separates intrapreneurs from highly productive employees.

Innovators are a naturally curious bunch. They take the initiative to wander outside of their sphere of expertise. They do not need a manager to provide a map because they are guided by technologies and problems about which they care passionately.

Figure 11 depicts the two areas of initiative on which an intrapreneur typically focuses. The intrapreneur must take initiative in both areas; anything less decreases the chance for effective innovation.

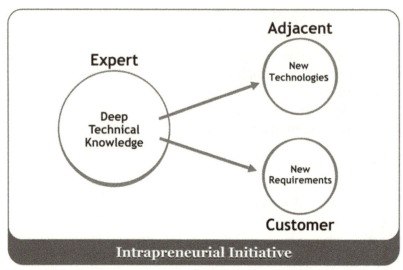

Figure 11. An intrapreneur builds on expertise by learning new technologies and customer requirements.

The first (and most important) area of initiative is the entrance into the customer sphere. It's important to be aware of the two classes of customer: external and internal.

External customers are the businesses or consumers that purchase the innovative, new high-tech products that are being built.

Internal customers are the individuals and teams that build, test, sell, and support high-tech products.

Regardless of whether the customer is internal or external, the intrapreneur must leave the comfort of his or her sphere of expertise and listen closely to customer requirements and problem statements.

Requirements gathering is a normal and mandatory part of any high-tech product development cycle. What is different about the intrapreneurial approach? The primary difference is that the intrapreneur is driven by personal interest. He or she is asking solely because of personal curiosity, or because of a passion for a particular field.

In the book *Innovate With Influence*, frequent intrapreneurial visits to the John F. Kennedy Presidential Library and Museum are made by an interested intrapreneur. Archivists at the JFK Library are using EMC equipment to digitally preserve physical copies of papers from JFK's presidency. These visits – essentially field trips – were not tied to any official product plan but resulted in multiple new ideas and areas of research for EMC. They were not controlled, legislated visits and were inspired by the intrapreneur's personal interests.

Voluntary, first-hand exposure to customer issues fuels the innovative fire. Problems or challenges witnessed at a customer site generate even more motivation to the person building the next-generation solution.

The second area of initiative is the continual drive to learn new technologies, especially those that are in some way adjacent to the intrapreneur's sphere of expertise. These self-guided investigations are most effective if they are driven by the exposure to customer problems and requirements. It is possible to explore new technologies outside of the context of customer requirements; this is called blue sky innovation. Customer requirements will often surface as the intrapreneur practices a third habit: collaboration.

Habit #3: Collaboration

Intrapreneurs are, by necessity, highly collaborative. The reason they practice Habit #2 (initiative) is because they fully recognize their need for new knowledge. They don't know all of the problems their current (or future) customers are experiencing and they lack comprehensive knowledge of an adjacent sphere of technology that just might light an innovative spark of how to solve a problem differently.

To overcome these limitations, intrapreneurs collaborate.

Innovate With Influence describes a collaborative effort to produce an idea that translated into multiple billions of dollars of product revenue. The idea was instrumental in giving rise to the adoption of disk array technology by mid-range corporations (one notch below enterprise customers in terms of size).

The idea was spurred by the need to solve the most important customer requirement of an information storage system: data integrity. In the 1980s, a new technology known as RAID had arisen. RAID used mathematical techniques to re-create information from failed disk drives. If the math was wrong, the customer would receive corrupt data. Customers wanted RAID because it was faster than any disk technology previously available.

Thousands of lines of new software had to be written to implement RAID. Dozens of failure permutations could impact RAID systems.

How could this software be tested to prove that the math never failed? The intrapreneur in this case was highly productive and an expert in RAID. He had taken the initiative to understand the customer sphere. Figure 12 shows a summary of the problem.

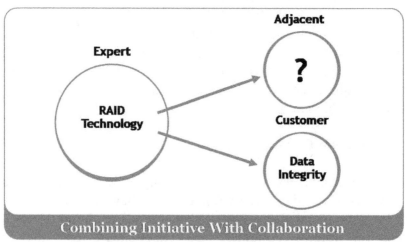

Figure 12. Customer collaboration identifies a problem (data integrity) that may be solved by finding an adjacent technology.

In this case, the adjacent technology was a very robust testing framework. This framework already verified data integrity, but it did not have a specific solution for verifying RAID disk arrays. Employees in the two technology spheres collaborated on new functionality that inserted tortuous fault events at every mathematical calculation point. This tool became internally known as the disk array qualifier (DAQ). The resulting disk array, known as CLARiiON®, achieved superior levels of quality that led to well over $12 billion of revenue generation in a 20-year period. Figure 13 depicts a Venn diagram of the DAQ solution.

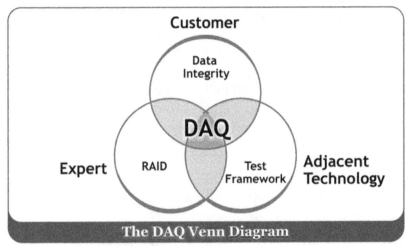

Figure 13. The Venn diagram technique is a classic approach to intrapreneurial innovation.

The three circles in Figure 13 represent the three basic habits of highly effective intrapreneurs. A highly productive employee becomes proficient in a particular expertise (in this case, RAID). The intrapreneur then takes the initiative to identify problems from the customer sphere (data integrity). Initiative spurs the search to find an applicable adjacent sphere (the test framework) and collaborate.

Collaboration between the two technologists yielded an innovative solution: the DAQ.

How can successful intrapreneurs discipline themselves to continually participate in collaboration? They develop the routine of practicing the fourth habit of successful intrapreneurs: individual 3-box time management.

Habit #4: The 3 Boxes

In Chapter 3, we discussed the corporate strategy of employee assignment to Vijay Govindarajan's three boxes of innovation. Corporations balance the distribution of R&D resources to three different boxes to meet present, past, and future needs.

The reasoning behind Vijay's corporate allocation of human capital can be transferred to an individual's allocation of time and lead to better time management. Figure 14 reviews the 3-box corporate strategy.

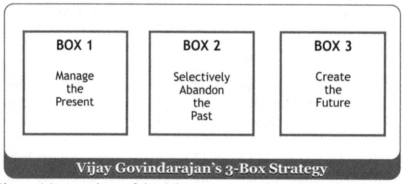

Figure 14. A review of the 3-box strategy of corporate investment.

Figure 15 applies this 3-box corporate framework to an intrapreneur's use of his or her own time (note that the box titles change when applied to an individual).

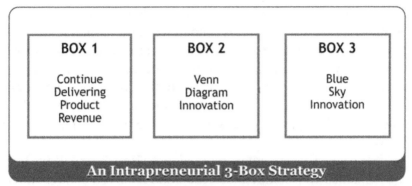

Figure 15. An intrapreneur's 3-box strategy allocates time to promote innovation.

Intrapreneurs can be most effective when they are delivering products as part of a business unit (as opposed to being a member of a research team in an ivory tower). Why? They need to be in the trenches, where they can be highly productive, visiting customers, and collaborating with others. They are respected within their organizations for doing those very things.

Perhaps their most significant contribution to their business unit's product line is funding their employment and that of their collaborators. They are squarely positioned in Box 1.

Spending all of their time in one area of expertise does not enable intrapreneurs to achieve success. Their natural curiosity and passion will not allow them to stay in only one place. They practice the discipline of limiting the amount of time they spend in Box 1.

By limiting the amount of time they spend in Box 1, intrapreneurs practice the discipline of regularly scheduled Box 2 and/or Box 3 activities. They set aside the time to learn about customer issues. They set aside the time to explore adjacent technologies. They regularly meet with experts in adjacent fields and collaborate to dream up ideas of what might be possible.

It is worth pointing out the difference between Box 2 and Box 3 intrapreneurial behavior. Box 2 behavior is characterized by Venn diagram innovation. The intrapreneur is collaborating in the context of a well-defined customer problem.

Box 3 behavior is characterized by blue sky innovation: taking the initiative to learn new technologies and collaborate without necessarily starting with the context of a defined customer problem. Blue sky innovators may ask themselves and others, "What might this capability be used to do?" Answers to this question can result in breakthrough innovation. It is often the case that breakthrough innovation can be applied to customer problems they don't yet know they have!

As Vijay Govindarajan pointed out, Box 2 and Box 3 allow large companies to "compete for the future." Even though it yields positive results, the fourth habit is very difficult to put into practice in large companies. The pressure for employees to spend 100 percent of their time in Box 1 is ever-present. Successful intrapreneurs, however, always find a way to make it happen. As a result of their clever approaches to time management, intrapreneurs are engaging to be around. They usually have multiple ideas on the back burner and make constant progress on all of them. Of primary importance is the fact that they continue to deliver (and usually exceed) their Box 1 goals.

Where do they find the time? What is the secret ingredient in their success? The answer to these questions can be found in the fifth habit: navigating visibility.

Habit #5: Navigating Visibility

Visibility avoidance – not actively publicizing one's talents – seems to be incongruent with climbing the corporate ladder. Doesn't everybody want to be recognized for their great ideas and promoted to higher levels of responsibility?

Intrapreneurs, quite simply, need not. Their intrinsic passion for the generation and delivery of ideas is their main reason for coming to work every day. They fill their time with trips to the lab to debug alongside their teammates. They close themselves inside remote conferences rooms and collaborate on whiteboards. They spend hours on their laptops, searching high and low for new sources of learning on the latest technologies. They visit local customers to check on their configurations.

They are not motivated by rewards and recognition from others. They are rewarded by the process that they have created for themselves.

This sort of self-motivation does not leave a lot of time for the bureaucratic meetings that are so common at large corporations. Intrapreneurs who find themselves lassoed in these types of meetings recognize they are jeopardizing the core value that is the hallmark of any intrapreneur: productivity. They understand that certain forms of recognition by executive management are accompanied by the unwelcome trappings of inefficient assignments and endless, countless meetings.

While they eschew recognition by executives, they strive for visibility with the builders in the trenches.

As a result of maintaining the delicate balance between maximizing trench visibility and minimizing corporate visibility, many intrapreneurs have adopted a "plus 2" approach to recognition within the corporation. They create extremely strong bonds of trust with their direct manager, reinforced by their consistently exceeding performance expectations, and strive for an equally strong bond with their manager's manager.

Any regular interaction with managers and executives higher than two levels above their current position is generally avoided. Building trust with the first two higher links in the management chain can accomplish two things:

1. It allows the management chain to run interference for the intrapreneur, which keeps the intrapreneur out of inefficient meetings.
2. It allows the management chain to advocate on behalf of the intrapreneur, which can advance those causes that require resource allocation outside of the business unit.

Why would the links on the plus-2 management chain go to these lengths to advance the cause of an intrapreneur? It comes down, once again, to the core trait of an intrapreneur: productivity. Corporate intrapreneurs have a track record of success. Their previous inventions became products that continue to generate revenue. People like to work with them, and they raise morale. Their management chain wants to keep them happy, and will typically do their best to eliminate corporate obstacles (or raise them to protect the intrapreneur's time).

Links on the plus-2 management chain rely increasingly on the sixth trait of a highly effective intrapreneur: building bridges to the future.

Habit #6: Bridge Building

Everybody who has ideas wants the luxury of time and freedom to tinker with them. Some people who generate ideas like to work on prototypes of their ideas. Few people have the capacity to work in Box 1 while advancing new research; those who do are the true intrapreneurs.

Anyone who develops prototypes of their ideas likes to have those prototypes analyzed and chosen for funding. The ability to pull off this type of work while simultaneously delivering on Box 1 commitments is one of the hardest balancing acts of all.

Yet intrapreneurs find a way to keep their ideas alive using a variety of different techniques. Figure 16 depicts the value of this type of behavior.

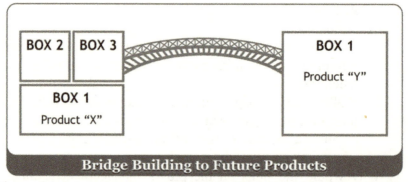

Figure 16. Intrapreneurs use Box 2 and Box 3 to drive new product offerings into the next Box 1.

Intrapreneurs distribute their time among the three boxes on the left hand side of Figure 16. They are delivering Product X while simultaneously exploring new ideas and problems. They use the results of the work in Box 2 and Box 3 to build a bridge to a new product offering (Product Y in what becomes their next Box 1). Building the bridge takes a lot of work and a lot of time. It also may take a lot of people, and these people are typically busy with Box 1 activities of their own.

How, then, do intrapreneurs build the bridge? The answer lies in their unique ability to keep ideas alive by inspiring and then leveraging human capital. Intrapreneurs commonly leverage human capital from one of four areas:

1. Interns. College interns are a wonderful thing. They often do not show up in a business unit's headcount. They are easier to hire than full-time employees, especially given the fact that they may only be available for a few short months. They are typically eager and productive. It is often less risky to have them work on something new (as opposed to relying heavily on them in Box 1). Interns are a great way to quickly build innovative new ideas, and the enthusiasm innate in intrapreneurs can be highly motivational for a college intern.

2. New employee training. New employees are often assigned to Box 1 activities. It typically takes new employees a while to get up to speed. Savvy intrapreneurs will take them in and help them to cut their teeth on some of the newer ideas. When introduced to Box 2 and Box 3 activities, the new hire may begin to mimic intrapreneurial behavior when they are ultimately moved into their own Box 1 job.

3. Outside of the business unit. Intrapreneurs are, by their very nature, collaborative. They are keenly aware of the work going on in other business units. They are quick to recognize slack in other organizations and present those

employees with the opportunity to contribute to new ideas. These collaborations, of course, often involve engagements between two different intrapreneurs from separate business units. A well-connected intrapreneur will often look to pull in resources from field engineers, too. Anyone who looks even marginally underutilized is fair game.

4. Pitching co-ownership. Most people enjoy being a part of a successful endeavor. An intrapreneur motivates employees within their own business units by articulating the value of Product Y. Co-workers quickly realize the contributions made by intrapreneurs and recognize that by dedicating part of their own time to Box 2 and Box 3 in support of "their" intrapreneurs, they will be lining themselves up for the best work if and when Product Y is approved. If they participate in the beginnings of Product Y, they become known as the best candidates to kick off the formal development.

By applying the tenets of bridge building, the intrapreneur creates a volunteer team that, together, turns valuable ideas into marketable products. In the process, the intrapreneur serves as a mentor to many. The end result is a Product Y proposal that usually has the teeth and the legs to be adopted as a new direction in the organization, given that the intrapreneur has a strong track record on the seventh and final habit: they finish.

Habit #7: Finish

Many organizations have dedicated research teams that are not tied to a shipping product. Business units may have advanced development teams whose members are able to propose next-generation technologies because they are not directly part of a team that is delivering a shipping product.

Some of the most effective intrapreneurs do not belong to either type of group. They are still in the trenches, often striving to deliver the ideas that they themselves helped to generate. They are finishing Box 1 commitments while starting Box 2 and Box 3 initiatives.

Their reputation as a finisher is what gives them the respect of the organization. They don't propose pie-in-the-sky solutions or architectures and then leave the team to figure it out. They deliver customer architecture: architecture that is realizable and deliverable to the market.

Intrapreneurs finish Product X by delivering it to market. They stay involved until the job is done. The trait of finishing closes a circle that connects with the first trait of productivity. Intrapreneurs finish because they are productive; it is, after all, hardly productive to use the company's many resources and have nothing to show. Figure 17 depicts the seven habits of a highly effective intrapreneur.

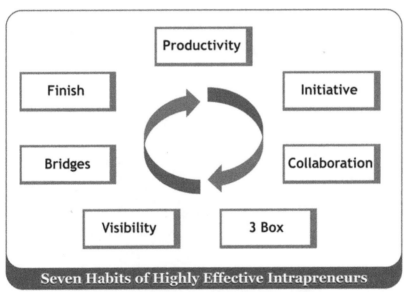

Figure 17. An intrapreneur's seven habits yield a continuous stream of innovation.

Put into action, these seven habits form the framework by which a corporate intrapreneur operates. Entrepreneurs at startups spend a good deal of time chasing funding; intrapreneurs at corporations spend a good deal of time chasing adjacent technologies and collaborative partners.

There is a strong relationship, therefore, between an intrapreneur and the corporate technology portfolio. When a corporation has an abundant and closely related portfolio of products and supporting competencies, intrapreneurs can thrive.

One corporation known for having a tightly integrated set of related technologies has also been called the fourth horseman of the Internet: EMC.

5. The Fourth Horseman

The search for an adjacent sphere of technology is a distinguishing trait of the intrapreneur. Once intrapreneurs have established their foundational career base of outstanding productivity, they differentiate themselves from other productive employees by taking the initiative to explore areas beyond their expertise. Large corporations can facilitate or hinder these initiatives. When handled properly, corporate enablement of the intrapreneurial spirit can result in accelerated innovation.

The stereotypical innovator is an entrepreneur at a startup or small company. These innovators are often hindered by lack of funds. In contrast, well-funded revenue-generating innovation can have far greater impact when global intrapreneurs effectively collaborate via adjacent spheres of technology. Keep in mind that these intrapreneurs have access and visibility to the needs of global markets. Subsequent chapters of this book will explain how best to weave together these global resources.

This chapter takes a deep look at the breadth of technologies available at a large corporation. It describes in detail how the technologies in a corporate portfolio can be combined and enhanced in ways that are not possible at a startup. Considered as a whole, the realm of corporate technologies becomes a playground of innovation for intrapreneurs, especially when their employing organizations have a strong central theme.

One strong central theme that is common at high-tech corporations is digital information. The explosive growth of the Internet in the 1990s was, in essence, a coming together of technologies that enabled the generation, organization, transfer, and storage of digital information. The press identified four companies that paved the way with skyrocketing sales of high-tech Internet products: Sun, Oracle, Cisco, and EMC.

These companies were often referred to as the four horsemen of the Internet. Sun provided the servers and applications that generated information. Oracle provided the database technology that organized digital information. Cisco provided the networking infrastructure that allowed information to be exchanged. And EMC provided the storage technologies that actually served as the final resting point for the bits and bytes of digital information.

The fourth horseman, EMC, was well known to many investors in the 1990s. It was the top growth stock of the decade, with a 10-year return of over 87,000 percent!

While investors were enthusiastically aware of EMC, the organization was largely unknown to the general public. Internet users and customers knew about Sun. Sun's servers, applications, and Java technologies were (visually) front and center. They knew that Cisco was providing the global connectivity to Web sites around the globe. They also knew that customer data was often organized in databases provided by Oracle. Users didn't knowingly interact with storage systems at the far end of the technology spectrum. Large EMC cabinets filled with spinning disk drives silently stored the world's data, out of sight and out of mind.

A decade later, information storage companies like EMC stood front and center. The world generates overwhelming amounts of digital information. Hundreds of millions of individual consumers generate terabytes of personal information. Researchers generate and analyze petabytes of scientific data. E-commerce and governmental infrastructures increasingly rely on large cabinets filled with digital storage devices to carry out their business-critical processes. The overall amount of information in the digital universe is now measured in zettabytes.

The storage and management of digital information is arguably one of the most difficult high-tech problems for any business to solve. Financially challenged organizations have less money to buy more disk capacity. The generation of electricity to power ever-growing numbers of disk drives is a financial (and environmental) concern. More and faster CPUs continually push the boundaries of performance for disk drive technology.

And many of these problems and technologies originally thought to be the exclusive domain of the wealthy nations are beginning to penetrate deeply into developing countries.

In other words, the global high-tech information industry is ripe for worldwide innovation. To fully understand the innovative capacity of a large high-tech corporation, it is worth taking a deeper look at a portion of the corporate technology portfolio of a company like EMC. This portfolio is the framework within which an intrapreneur must operate.

The essence of high-tech information storage technology is depicted in Figure 18.

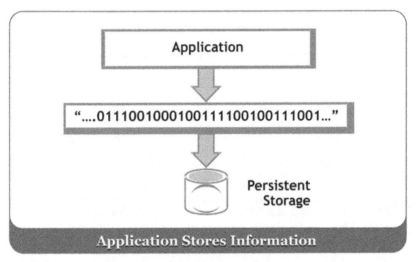

Figure 18. The core activity of the information industry is the generation and storage of digital bits.

Applications generate digital bits of information. The application could be an editor such as Microsoft Office Word. It could be a banking application that enables online financial transactions. It could be a video editing application that creates home movies. Whatever the application may be, the newly created digital information must be stored persistently. In other words, it must be remembered so that the application can retrieve it later, when you call for it.

Often times, corporations that are responsible for storing large amounts of digital information will purchase a disk array from a corporation such as EMC. Disk arrays are cabinets filled with disk drives. In the 1990s, EMC produced the highly successful Symmetrix® disk array, which powered information storage on the Internet. Figure 19 depicts the Symmetrix disk array functioning in its information storage role.

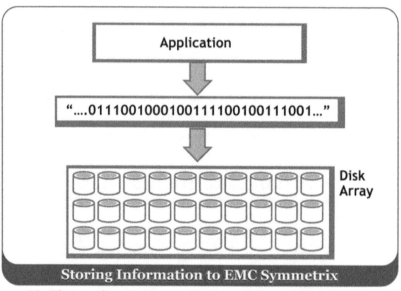

Figure 19. The need to store enormous amounts of digital information can be satisfied by a disk array.

Disk arrays such as Symmetrix contain complex software that allows for not only storage and retrieval of information but also for advanced features that provide additional value to an application.

In the early 1990s, an attempt was made to bomb the World Trade Center in New York City. Several businesses running within the World Trade Center were using Symmetrix to store their entire information portfolio. They asked EMC to add the capability to mirror the information to a remote geographic location (e.g., across the

Hudson River). EMC added a software feature known as Symmetrix Remote Data Facility (SRDF®). SRDF is depicted in Figure 20.

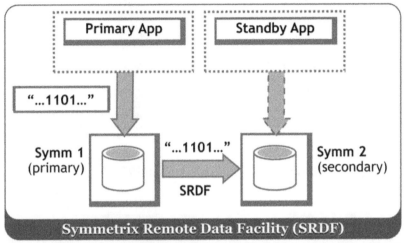

Figure 20. SRDF moves data from a primary site to a secondary disk array for added security.

This feature allowed an application running in a primary location (the World Trade Center) to have all of its information remotely protected. A standby application would be running at this location. This standby application would assume primary ownership of the information in the case of a disaster.

SRDF functionality preserved the critical digital bits of information required to allow businesses to keep running after the destruction of the World Trade Center's twin towers on September 11, 2001.

Businesses continued to request new features to supplement the original functionality of SRDF. In particular, a built-in capability to create point-in-time copies of digital information so that secondary applications could access the copy without risking corruption of the original data was keenly desired. Intrapreneurs at EMC accommodated this popular request. This feature, often known as snap copy, is depicted in Figure 21.

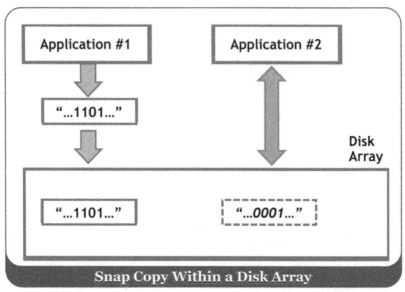

Figure 21. Snap copy technology allows a second application to access a point-in-time copy even when the original is modified.

As more and more businesses placed their indispensable information onto EMC equipment, they increasingly feared the possibility of downtime due to equipment failure. Each hour of downtime could translate into hundreds of thousands of dollars in lost revenue. Symmetrix already had the internal capability to recover from a single point of failure of any hardware component (e.g., disks or power supplies). This capability, however, did not provide full insurance for these businesses.

In particular, businesses were concerned that any failure in a singular path to a disk array could essentially cut off their application from their critical information. They began to set up multiple, redundant paths to the disk array to protect themselves from a single point of failure. On paper, the concept was reasonable, but the hardware configuration was not sufficient on its own. How would the applications know which paths were safe? Applications needed underlying software support that could discover, recognize, and utilize multiple paths to a disk array.

EMC provided a software technology known as PowerPath®. The technology could not only provide failover capability, but also execute load balancing for maximal performance. Figure 22 depicts the PowerPath solution.

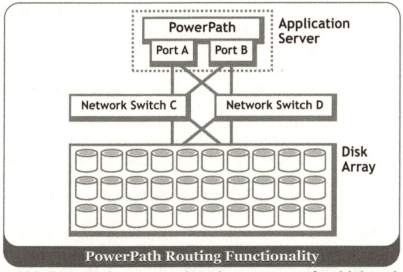

Figure 22. PowerPath can route data along any one of multiple paths if one path fails.

These technologies are but a few of the examples of why EMC became known as the fourth horseman of the Internet. On its own, the Symmetrix system dominated sales in large enterprise accounts. EMC's acquisition of a similar, mid-range disk array known as CLARiiON ensured EMC's continued leadership role. The two products, along with PowerPath, generated billions of dollars in revenue.

When the dot com bust occurred, EMC found itself in an unexpectedly vulnerable position. Demand for disk arrays declined. In the first few years of the new millennium, EMC embarked upon a strategy whereby it would expand and diversify its information technology portfolio via a systematic process of corporate acquisitions.

Corporate executives began searching for technologies that managed digital information *outside* of the disk array.

Documentum® was one such company and, in 2003, EMC purchased the company. Documentum technology managed the process of information workflow. For example, a mortgage application would result in the creation of digital bits that would enter a workflow process where the mortgage was analyzed, approved, granted, sold, and so on.

EMC also acquired Legato®. Legato provided technology whereby digital bits of information could be further protected by systematic backups to heterogeneous devices such as tape drives.

In 2004, EMC acquired VMWARE® technology, and moved closer to the applications that were generating the digital bits of information.

In 2005, EMC acquired Captiva®. Captiva provided information capture technology. When information crossed into an enterprise customer account (either electronically or on paper), Captiva would convert the information into a searchable digital format that could then be used as input into a document workflow technology such as Documentum.

The impressive pace of acquisitions continued. In 2006, EMC acquired information security experts RSA™. RSA is the world's most pervasively used encryption technology. In 2007, EMC moved into the consumer space with its acquisition of Mozy®. Mozy performs background scans of personal computers and transfers digital bits of information into a remote data center to protect consumers from PC crashes. In 2008, EMC acquired personal storage device maker Iomega®, and in 2009, EMC acquired information search technology expert Kazeon®.

Might an intrapreneur or a whole company of intrapreneurs have been able to generate these technologies? Given time, probably. But sometimes, it's quicker to acquire an existing good technology than try to reinvent the wheel without stepping on someone else's patents.

All told, EMC purchased over 40 companies in the first decade of the new millennium. Four days into the next decade (2010), EMC purchased information governance technology from a corporation known as Archer. Figure 23 depicts a subset of these 40 companies and the centrality they share: digital information.

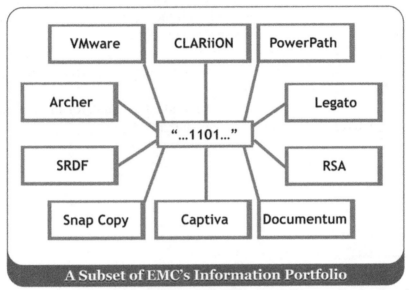

Figure 23. EMC's corporate technology portfolio revolves around digital information.

Corporate intrapreneurs need to develop a level of productivity and expertise in their assigned field. In the context of Figure 23, it becomes clear that EMC intrapreneurs need expertise in one of many categories that involve a specific aspect of handling digital bits of information. As they become exposed to customer problems, intrapreneurs can begin an in-house search of adjacent technologies that can result in an innovative solution to a customer problem. This in-house search is characterized by direct contact with intrapreneurs who are experts, not only in their field, but also perhaps in the entire industry.

This extremely rich ecosystem for innovation simply cannot be found at a startup.

The capstone of the ecosystem is the global distribution of technologists. EMC Symmetrix, for example, is supported by technologists in the United States, Russia, China, Ireland, India, and Israel. Each technologist has visibility to local customers and markets. This visibility is often inaccessible to their international counterparts. The local intrapreneur is presented with local problems, and they alone take responsibility for driving appropriate solutions.

If a corporation provides the bridges that allow local intrapreneurs to connect with each other globally, what results is a powerful, global engine of innovation.

The stories that follow are individual accounts of intrapreneurs who have taken the initiative to innovate globally.

SECTION II: GLOBAL INNOVATION

6. Shanghai: SYMSTAT Innovation

When students interview for jobs at EMC Corporation, they are often surprised to find that EMC is primarily a software company. EMC built its high-tech reputation in the 1990s through sales of the flagship Symmetrix hardware product. A Symmetrix cabinet is filled with hundreds of disk drives. The explosive growth of Symmetrix sales in the 1990s contributed to EMC's reputation as a hardware company (in spite of the fact that there is a large amount of intricate software running inside the Symmetrix disk array).

As mentioned previously, the attempted bombing of the World Trade Center in 1993 was an eye-opener for businesses and technology companies alike. Those that had been using a Symmetrix and storing their entire business information portfolio within the system quickly realized how vulnerable they were to disaster. If disaster struck, it could take a business days or weeks to acquire a new system, restore the lost data from tape backup, and attach the new system to a new set of servers.

Businesses turned to the SRDF solution to help them remotely mirror their data to a Symmetrix at a different geographical location (e.g., across the Hudson River). Customers would set up mirrored configurations known as disaster recovery sites. These mirrored locations would contain a second Symmetrix and a second set of servers. Should a disaster occur at the primary location, customers desired that the disaster recovery site would be up and running in minutes or hours (as compared to days or weeks with tape restore technology).

It was paramount that the remote Symmetrix device was an exact replica of the primary Symmetrix. The SRDF software engineers at EMC were tasked with the job of forwarding data written to the primary and safely mirroring that data to the remote setup. This type of replication is known as synchronous replication. With synchronous replication, the primary server receives an acknowledgement after the data has been safely placed on the secondary Symmetrix. Figure 24 reviews the SRDF solution.

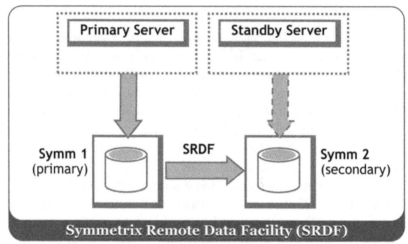

Figure 24. SRDF protects information by sending it
to a remote location.

The SRDF software was widely adopted in the industry. The Internet's explosive growth in the 1990s was due, in part, to customers' increasing comfort with e-commerce Web sites that were available 24 hours per day. Disaster recovery became an increasingly important business need; every hour that a business was offline could result in hundreds of thousands of dollars in lost revenue.

Initial deployments of SRDF solutions usually involved two Symmetrix systems (one primary and one secondary) and they were usually deployed for the sole purpose of recovering from a severe disaster at the primary site. As with any technology, however, customers soon asked for advanced SRDF features and created more complex configurations that addressed a variety of different business needs.

For example, some customers did not need the secondary system to be an *exact* replica of the primary. They only wanted to know that the secondary would be updated *eventually*. These customers wanted an asynchronous, ordered version of SRDF. An asynchronous version of SRDF would store new data in the large data cache on the primary Symmetrix, notify the primary server that the operation was complete, and send the new data (in sequence) to the remote Symmetrix at a later time.

This type of asynchronous replication (known as SRDF/A) is depicted in Figure 25. SRDF/A relies heavily on the size of the data cache and the speed of the connection between the primary and secondary Symmetrix systems.

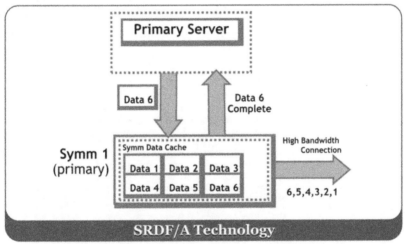

Figure 25. SRDF/A stores data in a large cache and replicates it (in sequence) to the secondary Symmetrix later.

In time, customers sought to upgrade, replace, or consolidate their Symmetrix systems. For example, after several years of operation, a customer might wish to upgrade to faster, larger, and newer hardware. In these cases, the customer would replace their old system with a newer system that had an appropriate amount of data cache capacity and a high-bandwidth connection to a remote system.

EMC would often provide the customer with a detailed analysis of their current cache size and available bandwidth. This report would be used to help the customer identify (and order) the correct hardware configuration for their future needs. This process involved some manual steps. A large report known as SYMSTAT documented all of the information about the current configuration and recommended another based on the anticipated load requirements. A SYMSTAT file contained thousands of lines of information, but the information was fairly straightforward to gather when one primary system was replicating to one secondary system.

Technologies – and customers' needs – are always evolving. Some customers opted for complex configurations involving multiple primary Symmetrix systems and the use of SRDF/A to replicate to multiple secondary systems. This type of "many-to-many" SRDF/A relationships between primary and secondary Symmetrix systems is depicted in Figure 26.

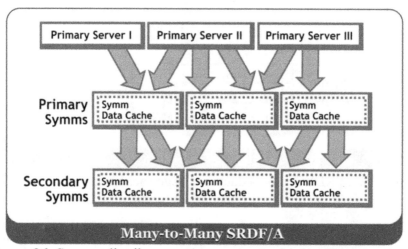

Figure 26. Servers distribute storage across multiple primary systems that replicate to multiple secondary systems.

When it was time for customers to upgrade these types of configurations to newer hardware, they looked to EMC for guidance. EMC field personnel would analyze these configurations using the same techniques as the more traditional setups. Unfortunately, the complex distribution of data and the multiple point-to-point connections between systems made it much more difficult to provide a fast and correct recommendation to the customer.

Peter Grogan is an EMC employee currently working in Hopkinton, Massachusetts. He is responsible not only for the quality of the overall SRDF solution, but also for the analysis of existing SRDF configurations and recommendations for the purchase of new Symmetrix systems.

As customers began deploying more and more complex SRDF configurations, Peter realized the amount of time required to analyze these deployments was increasing. He attributed this problem to the following factors:

- Servers were spreading data across multiple Symmetrix systems.
- The ratio of primary-to-secondary systems had grown from 1-to-1 to m-to-n.
- Different Symmetrix systems had differing amounts of memory and bandwidth characteristics.
- Eyeballing multiple reports from multiple systems was simply taking too long.

Peter and other employees were long-time, skilled, and veteran Symmetrix experts. They were struggling to find a quick and easy way to analyze mountains of information, generate a straightforward recommendation, and back up the recommendation with an easy-to-understand justification. In spite of their high levels of Symmetrix understanding, there was no ready solution.

Working on Peter's quality team in Shanghai, China, was Viki Zhang. Viki had only been with EMC for a year or two. She was an IT professional who had joined EMC with little knowledge of the information storage industry but a strong background in Microsoft™ Office™ Excel® technologies and Microsoft Visual Basic programming. Viki had proven herself to be a highly productive employee who consistently approached her assignments with great enthusiasm. Peter shared with her his customers' needs for accurate analysis graphs and the complex structure of the SYMSTAT file. He expressed his desire for an automated tool to quickly produce customer recommendations to match requirements. This problem is depicted in Figure 27.

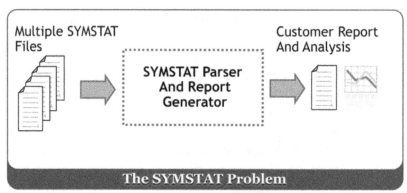

Figure 27. The SYMSTAT problem required new software to automate recommendations and analyses.

Viki knew right away that she could somehow apply her expertise to solve the customers' needs. Her situation can be represented using two-thirds of the Venn diagram innovation model. Viki was an expert in Microsoft Visual Basic® programming, and she also knew the clearly defined customer need for a detailed Symmetrix recommendation and graphical analysis report. Figure 28 depicts the initiative she needed to take to add an adjacent technology before she could set about creating a solution.

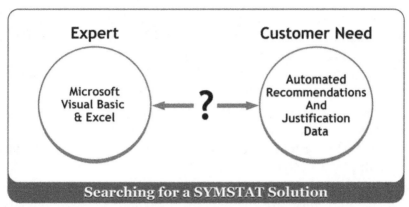

Figure 28. Adding an adjacent technology can help to solve a difficult problem.

Viki took the initiative to immerse herself in the Symmetrix technology. She learned about the Symmetrix cache. She learned about the different levels of Symmetrix bandwidth. She collaborated with Peter on the structure of the SYMSTAT file. She learned about consistency groups (groups of disks that function together to store application data).

She and Peter also began a series of prototypes. She wrote some software to analyze one small portion of a SYMSTAT file and generated Microsoft Office Excel spreadsheet results that could be converted easily to graphical information. Early results were promising and proved to her that she was on the right track. Her self-motivation to learn and her willingness to collaborate with her co-workers overseas had the potential to deliver a very important solution for the company.

The initial prototypes, however, were soon followed by myriad problems. Her software had to filter through thousands of lines of SYMSTAT data and recognize specific keywords. Parsing each file took time, and it was important to read the file efficiently (without making multiple passes). Her software had to input multiple SYMSTAT files and figure out which Symmetrix systems were primary, which were secondary, and what the replication relationships were. It needed to decipher the cache usage and bandwidth capabilities. The tool had to be extensible – who knows what other SYMSTAT keywords might end up being important over time? And finally, her software had to export data into Excel in such a way that report generation was extremely easy and 100 percent accurate.

All of this work was no small task for a person who had never even heard of Symmetrix before joining EMC!

After months of long nights and days of international communication, Viki displayed the most important characteristic of a global innovator: She finished. The SYMSTAT parser was able to input multiple SYMSTAT files and generate a detailed customer recommendation and analysis.

Her tool was brought into the field and exposed directly to EMC field engineers and customers. They ran real customer configurations through the tool and received accurate recommendations and analysis. Figure 29 depicts the final result of the SYMSTAT reporting tool.

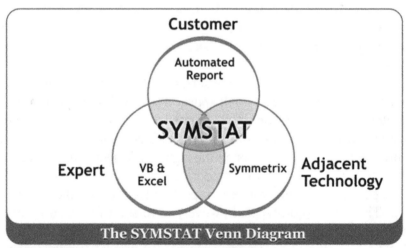

Figure 29. SYMSTAT innovation combined Symmetrix knowledge with Microsoft scripting and presentation tools.

The story of Viki Zhang's innovative solution displays many of the characteristics common to any intrapreneurial effort at a large company. First and foremost, she had already proven herself to be a productive, hard worker who had earned the right to innovate. Secondly, she took the initiative to understand a customer problem and learn about an adjacent technology (Symmetrix). Thirdly, she collaborated globally with her co-workers. These actions enabled her to develop a new expertise that she could combine with her already impressive set of skills. And finally, she finished the job, no matter how many obstacles she encountered along the way.

7. Beijing: Personalized Clouds

Customers who build their own data centers work with a variety of vendors. They buy servers from server vendors, network switches from networking vendors, and data storage devices from disk array vendors. Figure 30 depicts a typical data center integration that a customer will oversee.

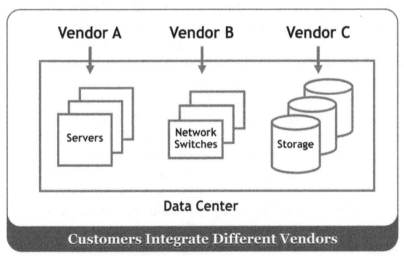

Figure 30. Customers who build their own data centers must build direct relationships with various vendors.

In the course of piecing together these technologies, they will receive personalized service from each vendor to assist in assembling their desired environment. The end result is often a data center that provides information technology (IT) services such as e-mail and order entry systems.

Often, the level of service is the reason a customer will choose a vendor. Customers build relationships with each of them. In turn, the vendor learns more about the customer, and most importantly, learns more about the best way to serve the customer's needs. As the relationship deepens, this in-depth knowledge of customer preferences can serve as a sales advantage when the customer decides to upgrade the vendor's equipment.

However, new paradigm shifts are threatening the traditional data center model (and therefore the traditional customer/vendor relationship).

Customers are beginning to consider the advantages of consolidating data center technologies; including participating in one-stop shopping for fully assembled systems (often called data center blocks or pods). These fully integrated systems eliminate the need for juggling multiple vendor relationships; instead customers can focus on one vendor relationship (the vendor of the fully integrated pod).

This approach is viewed by some as a temporary stop on the path to a complete and total outsourcing of IT services to a remote provider. In complete outsourcing situations, the server, network, and data storage choices are completely transparent and invisible to customers. These offsite and fully virtualized data centers are often generically referred to as "clouds." Within the cloud are a collection of CPUs that can run applications (compute clouds), as well as a collection of disk array systems for storing information (storage clouds).

Customers are beginning to build relationships with cloud service providers instead of traditional hardware vendors. The challenge for traditional hardware vendors is to become the hardware vendor of choice within the cloud. How can any given vendor maintain a direct relationship with customers in this new paradigm? How can they provide value-added services? Figure 31 highlights the separation between traditional data center customers and the vendors with whom they used to interact.

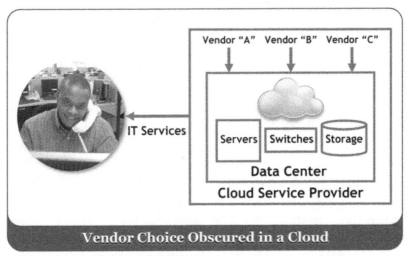

Figure 31. Customers are isolated from choice of vendor because these vendors interact only with cloud service providers.

In the new paradigm, customers work with cloud service providers instead of the vendors. Vendors work directly with the cloud service providers and provide solutions that integrate well and are easy to manage. This is a sensible approach.

Even more innovative would be an approach that allows vendors to continue building relationships with the customers of the cloud. If vendors could continue to offer significant relational advantage to customers, these customers would naturally choose a cloud provider that made use of its favored vendor(s)' products. It is incumbent upon vendors, therefore, to maintain customer connections while they establish new relationships with cloud service providers.

In other words, vendors like EMC would benefit from converting their face-to-face, personal relationships with customers into a set of personalized software services that extend beyond the cloud boundary and into the customer experience.

Hang Guo joined EMC's Beijing research team in July 2008. As a Ph.D. student, Hang sought to learn how to take ideas from the research phase and apply them to product implementations. In this sense, he wished to develop intrapreneurial skills and make a global impact on his new employer.

Hang did what any new intrapreneur on a team would do: he dedicated himself to productivity. He recognized that new ideas gain better acceptance within a large organization when they are presented by a productive employee with an excellent track record. In this spirit, he began establishing his local influence first.

Hang was assigned to work on a technology known as personalized digital search. Companies such as Google and Microsoft have well-known search engine technologies that run on individual desktops using keyword search. These technologies, however, can often be ineffective on desktops. The amount of information stored on personal computers is so enormous that users may have difficulty recalling the correct phrases or keywords that can quickly identify the correct document or file. The technologies that perform Internet search often yield less effective results on personalized searches.

Jidong Chen is a senior member of EMC's Beijing facility. He and his team spent a good deal of time exploring the problem of why Internet search techniques were less effective on searches for personal information. They recognized that Internet search leverages human intelligence. Search engines track human intelligence by measuring links between documents (page rank) and individual browsing behavior. The collective summary of this intelligence results in an effective search. More often than not, the correct document (or an appropriate document) is found.

This type of approach clearly will not be as effective in a personalized environment with only one user! Jidong and his team began to search for another approach. They started the process by asking themselves some important questions.

Can the concepts of links and page ranks be applied in a more unique way? Can we perform personalized desktop searches by leveraging human intelligence differently?

They decided to focus on human memory modeling. Memory fragments in the brain are connected by "memory cues" of user activity. When trying to recall a detail, many people will ask themselves, *"What was I doing at the time?"*

For example, consider the user who browses a particular Web page while writing a technical document on personalized search. The user stores a fragment of that browsing experience in human memory. Months later, the user desires to find that Web page again but is unable to remember appropriate keywords, but they do remember what they were doing (writing a technical report), and perhaps even when they were doing it. These connections are often remembered more easily than keywords.

Jidong's team brainstormed different ways to transfer the techniques found in human memory associations into a software solution. Their theory was that effective desktop searches could be accomplished by modeling associative memory techniques. Desktop links between documents could be formed by observing user activity. These links could then assist more effectively in a personalized search via a way that is highly relevant to a specific user. Figure 32 depicts this theory. User memory associations are captured as a set of semantic links representing activity between documents.

Associative Memory-based Desktop Search

Figure 32. Desktop users can find content more effectively using associative memory techniques.

Beijing: Personalized Clouds

Hang Guo joined Jidong's team and spent the next 6 months building an associative memory prototype known as My Memory Echo (iMecho). The iMecho prototype tracked user actions and stored associative links alongside desktop content. A new user interface was created that allowed users to enter keywords and "navigate their memory" to find the correct content.

Hang and Jidong worked closely with researchers at Fudan University in Shanghai and jointly wrote a paper describing their research and prototype. The iMecho idea was so compelling that their paper was accepted at the Association for Computing Machinery's Special Interest Group on Management of Data (ACM SIGMOD) conference in 2009. It was only the eighth time in this prestigious conference's 34-year history that a paper written in mainland China was accepted.

After 6 months of productivity, Hang began to propose his own ideas. He added a feature known as "context-aware search" into the next version of iMecho. At the end of his first year of employment at EMC, he had educated himself on the organization's involvement with IT clouds. He learned about the situations that customers experience regarding separation from cloud vendors. He started to wonder whether vendor information systems could be personalized. He took the initiative to understand the growing market for cloud service providers and identified three basic problems:

- Cloud service providers collect the logs from vendor hardware and are typically unwilling to share this information.
- End users of cloud services are reluctant to part with their personal information, and are often opposed to personal information being collected by service providers.
- Service providers are often not willing to invest the extra resources to capture freely provided user information and build complex models to leverage it.

Hang's involvement with iMecho had allowed him to become an expert in personalized desktop search. He had discovered a customer problem involving personalized cloud services. His realization is depicted using Venn diagram innovation notation in Figure 33.

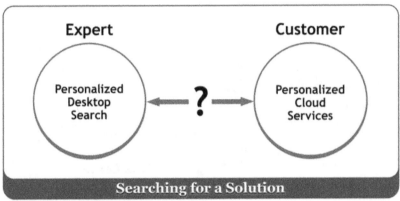

Figure 33. Can personalized vendor services and value be delivered from the cloud?

EMC has an internal business unit known as Digital Echo (Decho). This business unit delivers a cloud service known as Mozy. Mozy enables the backup of user information from a personal computer into a cloud. The details of Mozy's hardware vendors are invisible to the end user. The user identifies which directories they want backed up, and Mozy client software backs up the files into the cloud.

Mozy cloud-based services are an adjacent sphere of technology containing many different types of software modules. Hang recognized that while a personalization software module would be beneficial within the Mozy architecture, it would be much more feasible to move it down to the client of the cloud.

Then he realized he already had a client-side personalization module: iMecho.

The iMecho software keeps track of user activities and generates semantic and associative links. This information serves as an outstanding description of user preferences and behaviors. If the user had the ability to take some form of iMecho output and allow it to be securely accessed by vendor systems within the cloud, the bridge to providing user-specific services would be reestablished. Figure 34 depicts the transfer of iMecho semantic links into the vendor environment. In return, enhanced, user-specific services are provided.

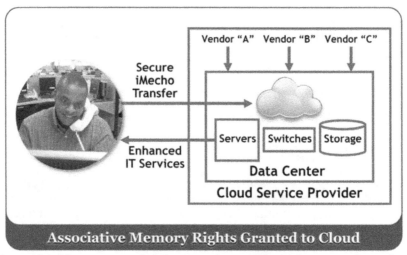

Figure 34. User grants cloud access to personal history and receives personalized IT services in return.

This solution benefits customers in two ways. Firstly, they need not keep track of activity and preferences (iMecho does this for them). Secondly, they are in full control of what the provider can see and they can choose to give as much (or as little) as they are willing to share.

The solution also benefits the cloud service providers and/or their internal vendors. iMecho output not only allows for instant user identification, but also provides personalized insight into what they are doing (and how they are doing it).

Hang's ideas were met with enthusiasm within EMC, and he began to share his ideas externally. Hang and his teammates prepared and published a second paper: "Personalization as a Service: the Architecture and a Case Study," and submitted it to the first international conference on cloud databases. He was invited to present his material at a workshop in Hong Kong. The experience allowed him to connect with his peers at other corporations (including IBM) and he continues to work with product groups within EMC.

Hang's idea of associative memory rights is another great example of Venn diagram innovation. He took his expertise (iMecho), found a customer problem (personalized cloud services), and searched for an adjacent area of technology (Mozy) that helped him to form a new and unique solution. The solution, known as Personalized Cloud Services, is depicted in Figure 35.

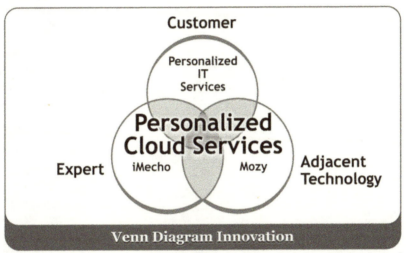

Figure 35. Classic innovation applies combined technologies to solve customer problems.

Throughout the process of developing Personalized Cloud Services, Hang displayed textbook intrapreneurial behavior. He joined a new company and focused on productivity. He did not try to impose his ideas upon the organization immediately. Within a year, he had built his influence and become a highly respected member of the team. He took the initiative to search for customer problems. He took the initiative to search for adjacent technologies. And of course, he used his influence to collaborate effectively, not only with his own co-workers, but also with his peers in academic and industry circles.

Beijing: Personalized Clouds

The last step of innovation is to finish the job and enable delivery of an idea, and Hang is in the middle of this process. The transition of IT services to the cloud will evolve over many years, and as a result of Hang's intrapreneurial efforts, these clouds will likely have personalized user services.

8. Bangalore: ESI Innovation

How big are the security issues being faced by organizations as they deploy their information infrastructures?

From small businesses to large data centers, the list of threats is nearly infinite. It can include worms, viruses, trojans, phishing attacks, denial of service, and so on. External threats are only part of the problem. Internal threats can range from bad behavior by malicious employees to seemingly innocent usage of internal servers.

Savvy organizations have an SOC (security operations center) equipped with technology that delivers security services by directly monitoring and combating the myriad threats intended to steal, block, destroy, or otherwise corrupt the valuable information assets that serve as the lifeblood of the business.

One responsibility of the SOC is to create a security command center environment, complete with a set of computers dedicated to monitoring and analyzing the threats that occur at (and within) the boundaries of a corporation or small business.

Technology planners at startups or small businesses recommend establishing an SOC in the earliest phases of infrastructure design. Figure 36 depicts an SOC in a small business.

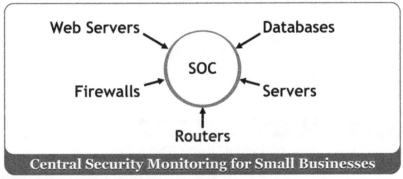

Figure 36. A security operations center can monitor all activities in a new information infrastructure.

A small business may have a Web server to interface with customers, internal servers that are used by employees, databases that contain customer information, and routers and firewalls to connect everything in a unified network. The SOC must continuously monitor this infrastructure, detect anomalies, attacks, and threats, and roll out corrective actions.

One of the more popular implementation choices for an SOC is the creation of a security information and event management system (SIEM). Logs and events from every device in the infrastructure are sent by the SIEM to a central monitoring location for analysis.

If the SOC has central visibility to all of the logs and events in the information infrastructure, events such as login failures, or incorrect passwords can be recognized and investigated. The SOC has access to the timestamps when these events occurred, and the devices on which these events occurred. It can detect intrusion patterns and learn to identify threats as they are happening with the objective of preventing them in the future.

One of the more popular implementations of an SIEM system is the RSA enVision platform (EMC acquired RSA in 2006). Figure 37 depicts an SOC managing security and compliance using RSA enVision.

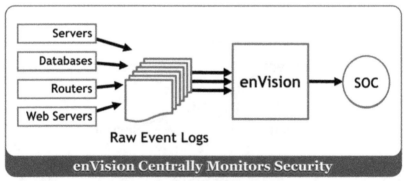

Figure 37. Example of an SOC using RSA enVision to analyze raw event logs.

Logs of events processed by any device or software in the information infrastructure are sent in real time to enVision. The administrator of the SOC can then perform analyses using enVision tools and receive reports about potential threats and vulnerabilities.

Arun Narayanaswamy is an intrapreneur and software engineer working on the enVision team in Bangalore, India. He and his co-workers have developed expertise in security technologies. As small businesses began deploying servers, databases, routers, and other software within their environment, Arun's team was assigned to automatically translate event logs into a format that enVision could understand. Figure 38 depicts the internal translation of raw event logs within enVision.

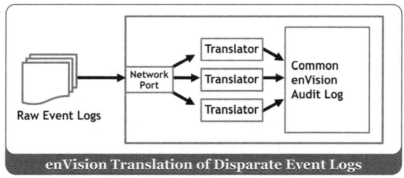

Figure 38. enVision translates disparate incoming event logs into a common internal audit log.

As part of their role in Bangalore, Arun and his team wrote translator software for the industry's most popular devices (e.g., Cisco routers and Microsoft Windows®-based servers). Every vendor's product required its own translator. The requests for new translators often came from corporate headquarters in the United States. The Indian team would study the set of events from a specific vendor, map the events onto a common set of security terms, and then embed this knowledge within the enVision device itself.

As the RSA enVision device grew in popularity, it became one of the more widely deployed security devices in the industry. Small businesses grew into larger enterprise deployments and the RSA enVision device would scale alongside the growing data center. SOCs that began with one administrator would often grow to a multi-person team. Data centers would add hundreds of new devices and countless different software packages in surprisingly little time.

A growing data center is a collection of different operating systems, databases, networking devices, storage systems, and applications (not to mention security software). When the growth is controlled, an effort is usually made to maintain continuity of brand and model. Often, though, someone decides to try something new and different. This deviation from the norm becomes a challenge for a security monitoring device such as enVision. Figure 39 depicts the challenge of handling event logs from various different vendors in a data center experiencing uncontrolled growth.

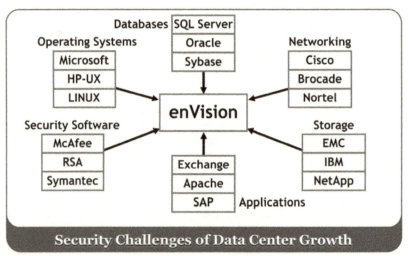

Figure 39. The cornucopia of hardware and software in a growing data center makes security monitoring a challenge.

It is not uncommon for large data centers to experience thousands of events per second. An unrecognized log entry is the equivalent of a security blind spot. A security administrator cannot evaluate whether an event should be perceived as a potential threat if no information is available. Blind spots are unacceptable events for customers, and there are numerous reasons why these blind spots could occur:

- Software upgrades for existing devices would contain new log events not recognized by enVision.
- Startup companies introduced new devices that were unknown to enVision.
- Internally developed devices were created by enterprise customers. These unmarketed devices had unknown log formats.
- Clandestine devices created by governments and security organizations did not share the format of their event logs with the general public.

Arun and his team were flooded with requests to fix blind spots, support new devices, and participate in professional service engagements for internally developed systems. Customers encountering blind spots could experience long response times as they waited for updated enVision software that could handle the blind spots they had reported; meanwhile, new blind spots would materialize. Competitors were nipping at the heels of the enVision team. Adding support for the latest event logs became a never-ending race to see which company could support the most devices.

Arun began a series of discussions with the executive management team in the United States. He quickly realized the expertise in his team simply could not scale to the level required to support dozens of new devices and thousands of new events. He had reached the point in the innovative process where his level of expertise was bumping up against a customer problem that required the assistance of an adjacent technology. Figure 40 depicts this familiar situation.

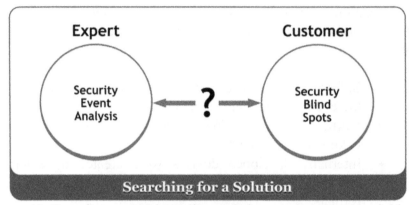

Figure 40. Security experts in India needed a new process to quickly fix security blind spots in customer environments.

The solution to this problem was less about introducing new technology and more about introducing a new process. Somehow, the team had to scale the process of preventing (or quickly fixing) security blind spots. Arun reasoned that the answer required taking the process his team used (writing internal enVision translators) and putting it into the hands of customers, partners, and vendors.

In other words, his team had to create a new tool: an event source integrator. This integrator would allow device vendors and partners to update enVision on their own.

Arun realized that this idea represented an opportunity to implement a solution completely outside of the United States. As an intrapreneur, he knew how to influence the team in his own geography. He began to communicate to them the following set of requirements:

- The tool must not expose the internal workings of enVision; this level of knowledge would be beyond the grasp of most users.
- The tool must not expose the common log format of the enVision device (XML). Not all users are familiar with XML technologies.
- The tool must be easy to use, intuitive, and graphical in nature.

- The tool must allow users to test the translation of device events into enVision events.
- The tool must allow for integration of new events into the enVision device.

The requirements presented by Arun required the creation of a highly usable compiler to simply translate input from one format to another. Arun assembled a team of developers and focused on two key activities: creation of a compiler and creation of a user interface.

The user interface would be a critical component of the solution. It would be completely developed outside of the United States and had to be exactly what was required by customers. The team developed the interface over the course of several 3-week sprints; the efforts of each 3-week sprint resulted in a user interface that could be shared with customers.

Dozens of potential users were provided with early copies of the software and feedback from each 3-week sprint's release was factored into the next 3-week sprint's activities. Over the course of several months, an interface was developed that met the needs of more and more customers.

When Arun determined the back end and front end of the software were ready to go, the team completed the first official release of their invention: the event source integrator (ESI). The ESI tool was given away for free to anyone who wished to plug the gaps of any blind spots in the enterprise.

Customers who had seen a new event go unrecognized could use the tool to map the unknown code into a recognizable event for enVision. They could immediately test and deploy this new knowledge into enVision without having to wait for a new release.

Device vendors with new event messages could parse them ahead of time and release their own enVision upgrades. Startup corporations that created new types of devices could announce their first product as being "enVision compatible."

What Arun and his team had done was find a way to transfer the most important aspects of their knowledge and expertise to a wider audience without requiring in-depth familiarity with enVision. Figure 41 depicts the Venn diagram for their innovation.

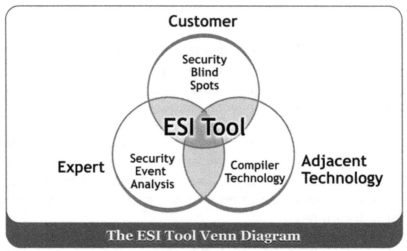

Figure 41. Development of the ESI tool was driven by customers' critical need to prevent security blind spots.

The ability of the team in India to finish the ESI tool from conception to delivery is a fundamental example of creating a global culture of innovation anywhere. In particular, Arun and his team showed the ability to build a bridge. They maintained their focus on enhancing the enVision product itself and simultaneously delivered a brand new innovation to solve a problem that could not be fixed using existing tools and processes.

9. Ireland: Certified Data Erasure

At some point along the timeline of information growth, one disk drive was simply no longer enough.

Disks have always been limited because they are mechanical devices that seek and spin. The time spent seeking and spinning caused disk drives to fall behind their CPU counterparts in performance. In the 1980s, the performance gap between CPUs and disk drives became so wide that the industry at large began looking for solutions. This problem (also discussed in the book *Innovate With Influence*), is depicted in Figure 42.

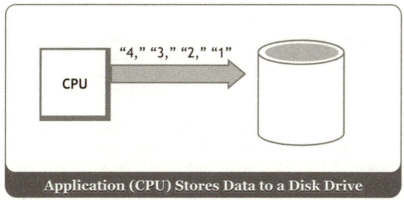

Figure 42. Fast CPUs flood slow disk drives with data, causing applications to wait for longer and longer periods of time.

The solution to this problem involved grouping disk drives together in arrays and inserting a cache between the CPU and the disks to accelerate the handling of reads and writes. The CPU continued to send multiple requests to what it believed to be one disk drive. In reality, these requests were being handled by a cache and multiple disks, and the CPU experienced much faster response times. This solution, known as redundant array of inexpensive disks (RAID), often included a parity disk. The parity disk contained extra data that could be used in the case of a disk failure (which was probable, given the increased number of disks). Figure 43 depicts this revolutionary improvement.

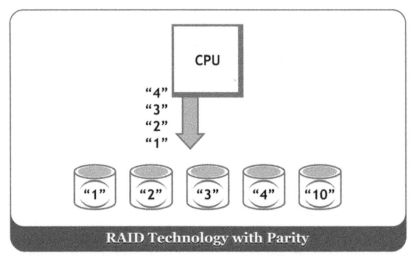

Figure 43. Data written by a CPU is handled in parallel by multiple disk drives, thus improving overall performance.

The advent of caching and RAID technology ushered in the era of the disk array. Factories began building large sheet metal enclosures populated with dozens of disk drives with built-in redundancy and fault tolerance. Each disk array contained redundant power supplies and cooling fans to go along with redundant cache and disks. As the disk array industry grew, the size of the disk arrays grew along with it. Eventually disk arrays would contain hundreds of disks.

One of the top EMC disk array manufacturing facilities in the world is located in Cork, Ireland. Millions of disk drives have arrived into Cork, been assembled into disk arrays, and shipped to customers worldwide. Over the years, the Cork facility's prestige in disk drive technology was unrivaled. Perhaps the most critical component of the engineering team's expertise was the relationships they fostered with each disk array customer.

Disk drives will eventually fail, no matter how expertly they are manufactured. EMC found themselves in the middle of a complex relationship between multiple disk drive vendors and thousands of customers. They developed sophisticated processes to track and handle disk failures at customer sites. The data gathered from customers would be forwarded to disk drive vendors to help them improve quality levels. This process was streamlined and fine-tuned over many decades.

Within the last several years, however, the changing landscape of high-tech governmental legislation would cause a major disruption to this process. Customers would use their disk arrays to store sensitive data. The leakage of any information found on these disk drives could be cause for steep fines and jail time for the offending parties. Figure 44 highlights some of the laws related to the protection of privacy of information, as well as the penalties to be imposed when these laws are broken.

Regulation	Potential Incarceration	Potential Fine
Sarbanes-Oxley	10 year	$ 15,000,000
SEC Rule 17a-4	Suspension	$ 1,000,000
Gramm-Leach-Biley	10 year	$ 1,000,000

Penalties for Failed Data Protection

Figure 44. Disk array customers faced stiff penalties for neglecting to protect information stored on disk drives.

Enactment of these pieces of legislation caused a major problem for customers returning failed disk drives. These disk drives were typically shipped to customers as part of a warranty program. This process allowed EMC to pass on initial savings to customers, provided that they returned the drives directly to EMC for root cause analysis.

The new regulations prohibited the shipment of disk drives back to EMC because the failed drives now contained sensitive information, which by law, could not be shared with a third party.

This situation hurt all parties involved, but it hit the customers especially hard. Drive failures could not be analyzed (which meant that the quality of disks was less likely to improve), and their inability to return disk drives affected the savings realized via the warranty program.

Were the drive to fall into the wrong hands during shipment, the customer could be immediately exposed. It would be easy for a government auditor to determine that a disk drive had indeed failed and question the customer as to the whereabouts of the disk drive. If the disk was unaccounted for, the customer could face the severe consequences highlighted in Figure 44. A solution to the problem was clearly needed.

A set of tools had been developed and used throughout EMC to overwrite every sector on all disk drives in an array multiple times (in compliance with government regulations). Unfortunately, there was no tool to electronically shred an individual disk drive. Customer engineer Shane Cowman took the initiative to work on a new tool that would meet governmental regulations, satisfy customers, and resume the flow of failed disk drives back into EMC. Figure 45 depicts two-thirds of the Venn diagram for creating an innovative solution to the problem of returning failed disk drives.

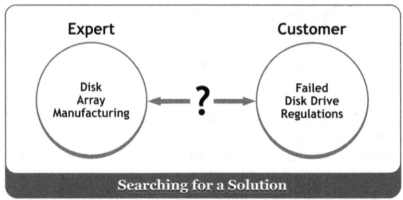

Figure 45. Customers were stuck holding on to failed disk drives.

As is often the case with new ideas, Shane and his co-workers began their own informal discussion of a solution. A quick study of disk drive failures at customer sites determined that over 70 percent of failed disk drives were still healthy enough to support the electronic shredding operations that would wipe the drive clean and allow it to be returned to EMC without risking exposure of confidential data. The high number of failed disk drives eligible to benefit from the solution represented a significant opportunity and the team decided to formally investigate.

Once the decision had been made to move forward with their investigation, the Cork team leveraged their strong bond with customers and the field. Requirements for a new tool were gathered from many different customers. During this process, it was discovered that field engineers for a customer in the Netherlands had already been experimenting with the idea of plugging disk drives into a personal computer and running a customized piece of software to erase information. The team analyzed the solution and found that it fell short of customer requirements in several different ways:

- The solution only worked on disk drives that had a fiber channel interface. Cork shipped other drive types as well (e.g., ATA and SATA).
- The solution only worked on one disk at a time. If the customer wished to erase multiple disk drives, the erasures had to be performed in a time-consuming, serial fashion.
- The user interface was not intuitive and was difficult to use.
- The equipment used to run the software was ad hoc. It wasn't scalable and the equipment didn't meet any of the stringent electrical certification standards routinely built into products manufactured by the Cork facility.
- The solution only supported disk drives with a sector format of 512 bytes. Cork shipped disk drives that supported 520-byte sectors as well.

- The solution had no mechanism to keep the drives cool during the erasure. This lack of cooling could actually do permanent damage to the disk drive.
- The solution did not conform to any standard. For example, the DOD 5220M erasure standard required a minimum of three overwrites using very specific bit patterns and algorithms.
- The solution did not produce a report that specifically recorded the identity of the disk drive and the result of the erasure attempt. This report would provide the critical proof of the erasure should the customer be audited.

Although the approach applied by the field engineers in the Netherlands was not the answer, evaluation of the process did provide the basic outline of a solution: Custom erasure software could be run on a dedicated PC that supported the insertion of different types of disk drives. Shane and the team began to match the long list of requirements with a set of hardware and software components that satisfied those requirements.

The first hardware component of the solution was a cabinet known as a disk array enclosure (DAE). The Cork facility had manufactured hundreds of thousands of DAEs over the years. One DAE could house up to 15 disk drives, and two of them could be connected together to house up to 30 disks. Figure 46 shows a hardware block diagram of a solution implemented using DAEs.

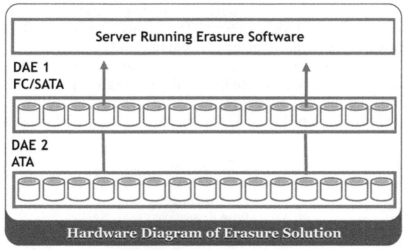

Figure 46. A data erasure solution could be implemented using DAEs.

The solution depicted in Figure 46 was elegant for a number of reasons:

- DAEs can hold more than one drive at a time, and the drives can be accessed in parallel.
- DAEs support three different drive types: FC, SATA, and ATA.
- DAEs can support future drive types as they are introduced.
- DAEs had the proper disk drive cooling mechanisms already in place.
- DAEs fit nicely into the racks already in place at many customer sites.

With the identification of the DAE as the foundation of the hardware solution, a surge of intrapreneurial activity took place in the Cork facility. The decision was made to work with an external vendor on the actual erasure software. The team collaborated with engineers in the United States on server hardware that would also possess the proper cooling and fit into the same rack as the DAEs. Documentation was developed for shipment with the product (as well as via download from the corporate Web site). Training was developed and offered via in-person instruction or via a webcast. The quality team verified that the hardware, software, and training protocols all worked as advertised.

In 2007, the Remote Change Management group even volunteered to service the product in the field as the primary line of support, and the first official certified data erasure tool appliance was shipped to customers. Figure 47 depicts the Venn diagram for this idea.

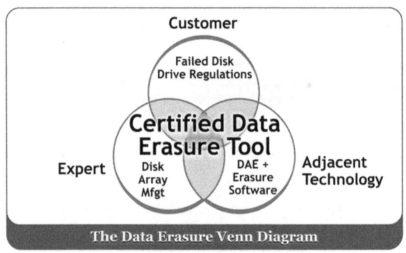

Figure 47. The Venn diagram for the certified data erasure tool.

Customer acceptance of EMC's certified data erasure tool proved so successful that a new group was formed in Cork to specifically target data erasures. Customers were pleased to resume their ability to return failed drives while adhering to the letter of the law and avoiding stiff penalties and fines. The tool continues to be updated as new drive types emerge; support for enterprise flash drives (EFDs) was added in 2008.

The team exhibited the classic behaviors of global intrapreneurs. They listened closely to customer needs. They collaborated globally with their co-workers. They designed a product that not only met all of the customer's needs, but also laid the groundwork for new drive types and erasure requests going forward.

They guided the success of the product from an informal conversation to a successful implementation. New versions of the tool continue to deliver value to global customers, many of whom must comply with different regulations in different countries.

In short, they did what all intrapreneurs must do.

They finished.

10. Israel: Component Collaboration

When it comes to innovation in the high-tech information industry, some countries have a more experienced engineering population than do others.

Perhaps no nation has more experience in the development of high-tech innovative ideas than Israel. For decades, Israel has produced some of the greatest entrepreneurs in the high-tech industry, and the area continues to be a hotbed of innovation for startup technologies.

One of the more complex problems within any customer data center is the need for continuous data protection and immediate (or nearly immediate) disaster recovery. The amount of time for which an enterprise can tolerate not having access to its critical information is growing shorter by the day. Nightly backups, in many cases, cannot be used to bring a business back online within minutes of a disaster because recently written data may be lost.

Most data centers contain a heterogeneous set of storage devices from multiple vendors. To safely protect the entire data center, it is often necessary to continuously replicate the data to a remote location. Depending on the region and the particular nature of the concern for security, the customer may choose to locate the remote site across town, across countries, or across continents. This type of replication at a distance is problematic when the data center contains storage devices from different companies because each device usually has its own way of performing remote replication.

To solve this problem, Israeli engineers developed vendor-neutral, continuous replication technology. This solution, currently known as RecoverPoint®, relies on a piece of software known as a splitter, and a piece of hardware known as a RecoverPoint Appliance (RPA). The splitter intercepts application data and sends it to the appliance. The appliance uses sophisticated algorithms to continuously replicate the data to a remote RPA. As a result, application data written just seconds before a disaster can be brought back online remarkably quickly with relatively little effort. Once the splitter has received an acknowledgement from the RPA, the data is then forwarded to the local vendor storage system. Figure 48 shows the overall architecture for this solution.

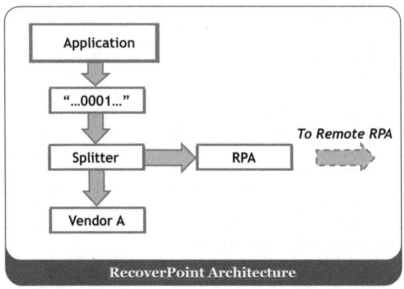

Figure 48. RecoverPoint technology provides continuous replication to remote locations.

The solution involving RecoverPoint and RPA provides a greater level of service than traditional backups (which are limited to a specific point in time). In addition to its remote capabilities, RecoverPoint can also provide continuous data protection within the bounds of a local data center. Whether the solution is local or remote, customers can recover from disasters or interruptions quickly and with minimal data loss.

Splitter technology is highly portable and implements continuous replication via interaction with the RecoverPoint appliance. The appliance accepts write operations from the splitter and communicates with a RecoverPoint appliance at the remote site. Figure 49 depicts two common deployments for the splitter.

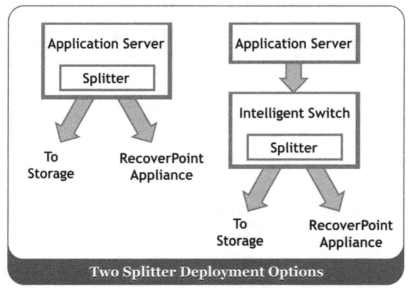

Figure 49. Splitter technology can be deployed within an application server or intelligent network switch.

The engineers at the Israeli COE were not content to perform incremental improvements to their product. Assaf Natanzon is a CTO and intrapreneur responsible for the RecoverPoint technology in EMC's Tel Aviv office. As the primary developers of RecoverPoint, he and his team had already proven to be productive employees. They also possessed the second and third habits of successful intrapreneurs: initiative and collaboration. These qualities led to an outpouring of global innovation that resulted in a multitude of new product releases.

As the popularity of RecoverPoint grew, the technology came to be deployed in increasingly large data center configurations. RecoverPoint is a very attractive option in facilities that employ disk arrays because the centralized storage devices are shared among multiple servers and applications. The connectivity concerns of large disk arrays make a utility such as RecoverPoint an important support mechanism for data integrity. Figure 50 depicts multiple applications consuming a centralized disk array.

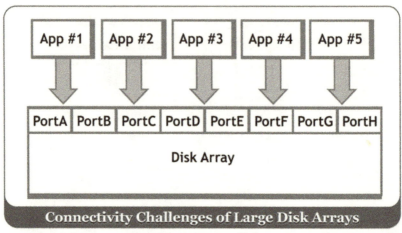

Figure 50. Multiple applications share a disk array, so data integrity is even more important.

Deployment of RecoverPoint technology in a configuration such as the one as represented in Figure 50 requires a customer to either install the splitter component on every server or install the splitter component on a layer of intelligent switches sandwiched between the applications and the disk array. In both cases, the customer will also deploy the RecoverPoint appliance.

Installing the splitter on every server must be done manually and, even though the process is not particularly difficult or time-consuming, if a large number of servers are involved, time is still money. The process time and commitment to both install the splitter and maintain its software revisions is significant. Installing the splitter on a layer of intelligent switches is also a manual process that introduces additional cost (and hardware) to the configuration. To address customers' concerns about time and expense, the RecoverPoint team collaborated with the CLARiiON engineering team, whose popular mid-range disk array storage device has hundreds of thousands of units deployed in the field.

The RecoverPoint team had already designed the splitter component to be deployed in multiple environments. The Israeli team brainstormed with the CLARiiON engineers on a solution that would satisfy customer needs by porting the splitter component into the disk array itself. The splitter could now be deployed in three different components: a server, a switch, or a disk array. The disk array solution provided the best of both worlds: no host footprint and no extra cost. Figure 51 depicts the RecoverPoint solution running within a CLARiiON disk array.

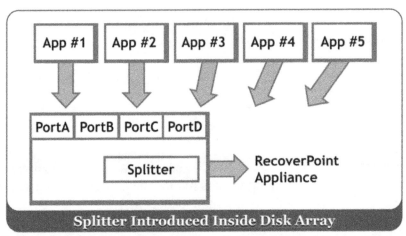

Figure 51. Splitter running within a disk array provides continuous protection with minimum customer configuration.

Key members of the CLARiiON team were located in various offices around the globe. Some idea exchanges were conducted by e-mail and phone, but some really needed to be conducted in person. By taking the initiative to travel overseas and collaborate with adjacent technologists, the developers at the Israel COE effectively influenced a separate organization and worked together to deliver a new product to the market. Customer feedback for this solution was overwhelmingly positive.

Customer satisfaction was not enough for this team, however. They were not done innovating.

The RecoverPoint Appliance (RPA) frequently sends data over long distances to a remote location. Assaf had developed data compression algorithms that ran inside the RPA and dramatically reduced the amount of data sent to a remote data center. Figure 52 depicts the value of these algorithms.

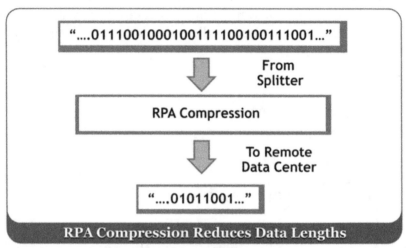

Figure 52. Compression algorithms reduce the amount of data sent by RPA to a remote data center.

Compression algorithms balance the trade-off between time (how long it takes to compress data) and efficiency (how much of a reduction in bit length occurs). Assaf started with an industry standard algorithm known as Lempel-Ziv encoding and modified it to satisfy the needs of RecoverPoint customers. By implementing compression algorithms within the RPA, Assaf was able to reduce the amount of time required to safely store customer data at a remote location.

During Assaf's travels outside of Israel, he frequently discussed his work regarding compression algorithms. He was convinced these algorithms could be componentized and used within other products. Assaf sought opportunities to innovate and soon began a series of simultaneous collaborations with multiple product teams.

The Symmetrix team already had a long history of successfully protecting customer data on remote sites via the SRDF product. Like Assaf, leaders of this team had also recognized the value of compression algorithms for reducing the amount of data transferred to the remote site. Symmetrix was already using a hardware compression option, but the team was in favor of investigating options that would allow them to deploy a less costly software solution with better compression ratios. The Symmetrix team was pleased to learn that Assaf had developed EMC-owned compression software as part of the RecoverPoint product. Even better, Assaf's software was portable.

The Symmetrix engineers began a formal evaluation of the RPA compression software. They discovered customer workloads shifted the balance between the time and efficiency of those algorithms. A back-and-forth collaboration between Assaf and the Symmetrix team ultimately resulted in a new piece of software being added into the next Symmetrix release: SRDF compression. Figure 53 depicts this solution.

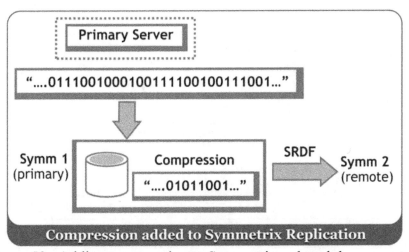

Figure 53 . Adding compression to Symmetrix reduced the amount of data sent to remote sites via SRDF software.

The collaborative efforts between Symmetrix and Assaf improved the software to the point at which the new compression algorithms were ported back into the original RPA. RecoverPoint customers benefited from the newer, faster compression algorithms that were also running inside of Symmetrix. Previously happy customers were now happier customers.

At the same time as he was working with the Symmetrix team, Assaf was also collaborating with yet another product group within EMC: the Viper team. The customer need for data reduction was an extremely hot topic. As IT budgets shrink the need to use existing storage more efficiently grows. The Viper team was working on a technology closely related to compression: data deduplication.

Deduplication technology (also known as dedup) searches incoming bit streams for duplicate bit patterns and only stores them one time (instead of multiple times). Figure 54 depicts a deduplication service that analyzes twelve different bit streams travelling between a server and a disk array.

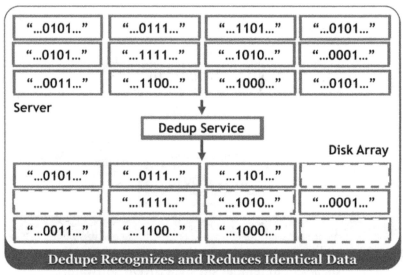

Figure 54. Deduplication software can detect identical bit patterns and store them one time only.

EMC's Viper team has the responsibility of creating portable deduplication software that can be deployed in multiple products. Assaf's compression software was closely related to the deduplication algorithms. In addition, Assaf had already proven that his algorithms could successfully be reused between two different products (RecoverPoint and Symmetrix). It seemed logical, then, that Assaf should collaborate with the Viper team.

Assaf began to assist the Viper team in the creation of a reusable software component that could provide a combination of deduplication and compression features. The software component would present an easy-to-use application programming interface (API) that would allow the software component to be embedded inside any number of EMC products.

Assaf visited the United States a few more times to collaborate with the Viper team and integrate the compression algorithms into the team's software component. The RecoverPoint technology innovation was being used for the benefit of yet another product team.

In addition to working with the Viper team in the United States, Assaf traveled to the Russia COE to learn more about Viper's deduplication algorithms (the Viper team was distributed across multiple geographies). He began to study whether or not deduplication could be integrated into the RecoverPoint product. The collaboration with the Viper team was becoming a mutually beneficial exchange.

During this innovation-sharing initiative Assaf exhibited all of the habits of a successful intrapreneur. He collaborated openly and assisted with testing and performance measurement. He reviewed documentation and participated in meetings between the Israel COE, the Russian COE, and the United States. In classic intrapreneurial form, he worked on these activities in parallel with his day-to-day duties on the RecoverPoint product, continuing to add features and enhance the product in line with customer needs.

Perhaps one of the most difficult tasks that an intrapreneur must accomplish is the bridge building from legacy products to new products. The success of the Viper compression and deduplication software collaboration would be measured by the successful launch of a product that contained the Viper software itself. In 2009, the Viper team's component was successfully productized within the Celerra® file server product. For years, the Celerra product has been one of the most popular file servers in the industry. Figure 55 depicts this solution.

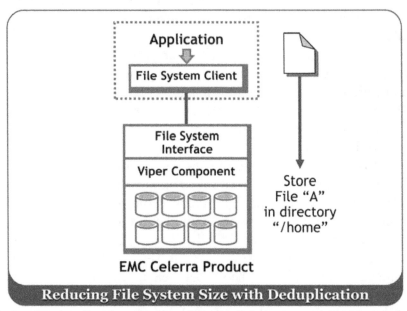

Figure 55. Applications storing files to Celerra automatically use less space via the Viper component.

If the file being stored in Figure 55 (file A) is a duplicate (i.e., it already exists somewhere else in the file system), then the compression algorithms present within the Viper component are used to shrink down that file as it's moved to another location. The RecoverPoint compression algorithms were now being used in yet another device.

Along with the compression of files in the Celerra product, the algorithms made their way into even more EMC products. The CLARiiON device (in addition to the already ported splitter component) added the compression algorithms to its feature set. This particular implementation compressed raw blocks of data (as opposed to the files being compressed in the Celerra product). In many cases, the use of compression results in data footprint reductions in the range of thirty to fifty percent!

The Networker® software within the EMC product portfolio performs data backup to multiple different storage devices (such as tape drives). For companies that do not want to manage the sometimes cumbersome process of working with tape drives, Networker offers an option of backing up data to an Internet-based cloud storage device. Assaf expanded his sphere of intrapreneurship even further by working with the Networker team to explain the benefits of compressing the backup data before sending it to the cloud storage device. This collaboration resulted in compression being added to yet another EMC product: Networker. As a result of data compression, much less data needs to be sent across a wide area network (WAN) and backups can be performed more quickly and consume less capacity.

Assaf and his team leveraged their intrapreneurial approach to solve multiple customer needs, including a few that customers didn't know they had. They collaborated with their global co-workers to help deliver new CLARiiON, Symmetrix, Celerra, and Networker functionality that made previously happy customers even happier. The engineering team in Israel continues to take the initiative and collaborate with other groups within EMC as they drive innovation into global markets.

11. Russia: Captiva Innovation

Information can come out of everywhere, and it can come out of nowhere. For a business owner, whether small or large, information capture is an enormous task.

Every business struggles to manage the flow of information into and out of their office. Whether it is mortgage applications, insurance claims, trade confirmations, or new account enrollments, consider the number of different ways that information can arrive into a business operation:

- Paper documents arrive through the mail.
- Fax machines spit out incoming documents.
- Scanners turn paper into electronic form.
- Employees receive business emails.
- Files are transferred directly onto corporate servers.
- Forms are filled out by customers via a corporate Web site.

The complete list is much longer. If these documents are paper, they'll often get filed away. In a large company, different divisions may use different filing techniques. If these documents are electronic, they will often be moved to a centralized data center for long-term storage. It takes a great deal of effort to organize and track every single document that arrives into a business environment. The cost is also quite high. Customers who choose to ignore information capture may find out their decision costs them more in the long run. They run the risk of failing an expensive audit.

Government agencies around the world have invited themselves to "stop by any time" and ask to see a specific piece of information. For example, the Securities and Exchange Commission can ask for a list of all e-mails related to the trading of a particular stock. The business that cannot comply with the request may incur significant fines or worse.

The problem of managing incoming information is further complicated by the fact that as documents arrive into an enterprise (in whatever form), they need to be routed and fed to applications that perform processing on the document content.

How can human beings efficiently receive, sort, route, store, and eventually find all of the documents in their enterprise? Increasingly, they are coming to rely on document capture software.

EMC Corporation has diversified its information storage portfolio by acquiring technologies that assist businesses in the capture of information. In 2005, the organization acquired Captiva, a company whose products transform business-critical paper, faxes, and electronic data sources into business-ready content. This content is suitable for processing by applications, and it is also ready for corporate audits. Figure 56 depicts Captiva's goal of information capture across a wide geography.

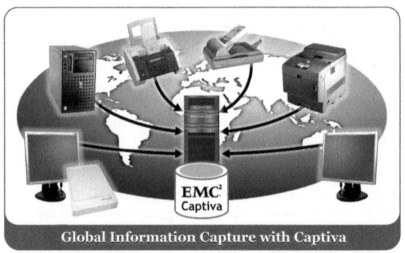

Figure 56. Global information capture from multiple types of devices.

The acquisition of Captiva allowed EMC to offer its customers the benefits of a paperless office:

- Reduce the amount of time spent handling, copying, storing, and finding paper.
- Reduce the amount of manual data entry required to extract information from paper documents.
- Improve the ability of multiple parties to collaborate on a single document quickly via electronic methods and workflows.
- Reduce the costs for printing, shipping, and storing paper documents.
- Improve the ability to find critical documents in response to legal discovery requests.

One of the key assets in EMC's Captiva portfolio is a piece of software known as InputAccel. InputAccel is the foundational software required for capturing incoming documents. The software captures, analyzes, recognizes, and indexes incoming documents and stores the results into a content management system (CMS). InputAccel supports many different types of CMS solutions, but has a long history of strong integration with the CMS supplied by Documentum (DCTM). Figure 57 depicts the high-level software flowchart for InputAccel integration with a DCTM CMS.

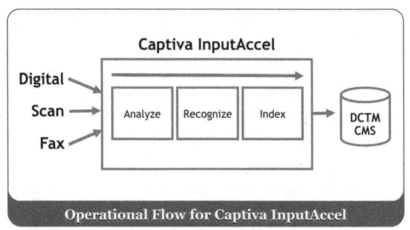

Figure 57. High-level operations performed by InputAccel
feed into a CMS.

In 2006, the Captiva software development team was located in two separate locations in California: Santa Clara and San Diego. As part of a strategy to increase the sales of document capture software in the Russian geography, the Captiva division decided to expand its development team into St. Petersburg, Russia.

Approximately 900 miles south of St. Petersburg is the Russian city of Belgorod. Belgorod is the hometown of software developer Denis Kiryaev. Denis had graduated with a degree in computer science from Belgorod State Technological University. Upon graduation, he began working on medical records management systems for a local software manufacturer. Given the growing high-tech job opportunities to be found in St. Petersburg, Denis submitted his resume to a St. Petersburg company that he had never heard of before: EMC.

Denis' background with medical records software was a natural fit for Captiva's new InputAccel team. He was hired and moved to St. Petersburg. He began learning the InputAccel software architecture. He also began learning about the Documentum CMS.

Two years before the acquisition of Captiva, in 2003, EMC had acquired the content management company known as Documentum. The software solutions provided by Documentum revolved around a digital repository for storing two things: documents and the information about them (also known as metadata). This digital repository is the CMS mentioned earlier. Documentum also created an application programmer's interface (API) that allowed third-party applications to store documents and metadata into the CMS. This API (called DFC) was Java-based and allowed remote client applications to store documents into the CMS, as depicted in Figure 58.

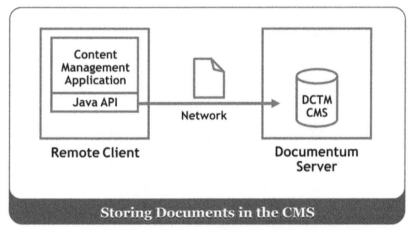

Figure 58. Java API stores documents in the CMS.

Not every application developer wishes to use a Java™ API to access the CMS. Microsoft Windows developers, for example, are often more comfortable using Microsoft-specific methods of storing content. In the 1990s, Microsoft developers preferred to use the Component Object Model (COM) to write their applications. In the late 1990s, Microsoft developers began to adopt the .NET programming API. The Documentum team made provisions for both of these programming styles by using layering techniques.

Figure 59 depicts two different software stacks that allowed Microsoft developers to store content into the CMS. On the left, the COM API is layered on top of Java. Legacy (older) applications are likely to use this approach. On the right, the .NET API is layered on top of the COM API. Newer applications (such as InputAccel) would use this layering option.

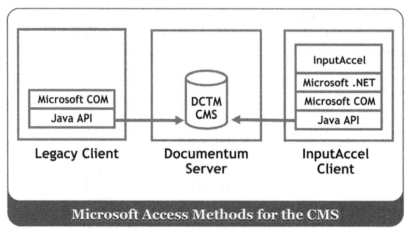

Figure 59. InputAccel's internal architecture uses Microsoft .NET programming techniques.

In 2008, as the St. Petersburg development team members began to ramp up their support for new InputAccel development, their Documentum co-workers announced that the COM layer would no longer be supported. COM was a 15-year-old Microsoft technology whose time had come and gone.

The decision to deprecate COM support had a significant ramification: The .NET bridge between InputAccel and the CMS was essentially cut off. Any new development on InputAccel would not be able to leverage any new features introduced into the CMS. The CMS was exclusively moving towards more Internet-friendly interfaces such as Java and Web services.

This left the InputAccel team with three (very costly) options:

- Completely rewrite the Documentum export module to use Java instead of .NET.

- Completely rewrite the Documentum export module to use a new Web services API instead of the Java-based DFC method.
- Write their own bridge from .NET to Java, or use Web services natively from .NET.

The first two choices were unrealistic. The InputAccel Documentum export module had taken years to build and ripping out the .NET interface was not an option.

The final choice also presented challenges. Two possibilities were examined. The .NET-to-Java bridge made the most sense because the Web services option would not perform as well. The team searched for an external .NET-to-Java solution. Only one was found, and it was too costly. It was determined that writing a .NET-Java bridge in-house would take two people 9 calendar months. This estimate did not include the amount of time it would take to thoroughly test this approach.

Customers needed the next version of InputAccel much, much sooner than any of these options would provide. A solution seemed impossible. There was simply too much new software that would have to be manually written and tested.

To compound the challenge, Denis Kiryaev had no experience with Java technology. His expertise was the .NET programming language. His team was faced with a seemingly unsolvable problem. Captiva's InputAccel customers depended on CMS integration via .NET. Denis had quite a bit of experience with automated code generation in the .NET framework. Figure 60 depicts the problem that he and his team were facing.

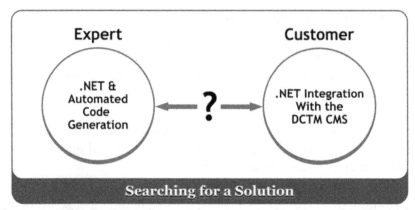

Figure 60. The St. Petersburg software team faced a problem without an apparent solution.

Denis, with his expertise in .NET programming and automated code generation, did not have a ready answer to solve the customer need. He was limited in his knowledge of Java, although he did know that Java had a native interface that allowed it to integrate with other languages. He was in the beginning stages of innovation. In a classic demonstration of intrapreneurship, he decided to reach out to one of his peers in an adjacent sphere of technology: the Java programming language. His colleague directed him towards the Java-native interface (JNI). Denis quickly learned that the JNI interface had almost everything he needed to integrate between .NET and Java, but it was extremely error-prone and complex. There was no way to quickly write such complicated software.

And then the light bulb went on.

Denis wondered if his experience with auto-generation of .NET code could be applied to Java. In particular, he wondered if he could write a piece of software that took the Java DFC API as input and produced wrappers on top of Java JNI that could be called by .NET. He was quickly consumed by the idea. In a classic example of 3-box behavior, he pursued his idea with a passion in parallel with his daily duties.

After a few days of experimentation, he ran the idea by his manager.

"You're crazy," replied his manager.

"I know," was Denis's response.

However, both of them realized that if the approach had even a glimmer of hope, it might save the team months of programming time. A computer can produce code infinitely faster than a human. And none of the InputAccel software would have to change. The solution had little to no impact on InputAccel itself.

Given that Denis had proven to be an extremely productive employee, his manager gave him 2 weeks to dedicate to further investigation. He dove headfirst into the effort. He quickly realized that his lack of Java experience was a problem. The allotted 2 weeks turned into 3, and 3 weeks turned into 4. Denis was running out of time, but as each week passed, he was convinced that it could work. Five weeks later, Denis was able to automatically generate code that replaced the COM-to-Java bridge.

Victory. He had done it.

His next step was to leverage the international relationship developed between St. Petersburg and the California teams. He knew that they would have the same reaction as his boss: "You're crazy."

The California team listened to his idea. Initially they thought the same thing: "This Russian guy is crazy!"

Denis had prepared himself for the conversation and showed that his solution was not only simple and feasible, but also met the schedule. He wrote a thorough specification to prove his point. His invention, the Java-.NET code generator, had reduced the development time from 18 person-months down to 3. Figure 61 depicts the path of innovation that Denis took to combine a customer problem with his own expertise and his new Java knowledge to solve a very difficult problem.

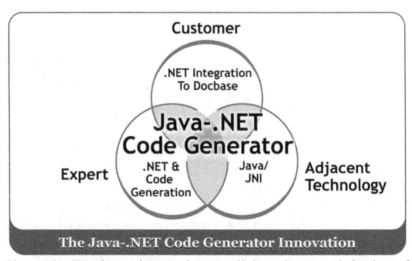

Figure 61. Employee innovation + collaboration = satisfaction of customer requirement.

Denis received approval to go ahead with his crazy idea and he completed the task. InputAccel is now fully capable of leveraging new CMS features, and all new versions of InputAccel will contain his invention. Customers who have already invested in the InputAccel technology will reap the benefit of his idea.

The story of Denis's invention follows the familiar pattern for any corporate intrapreneur. Once a track record of productivity has been established, intrapreneurs are given the freedom to explore (they don't need to ask permission). They find problems that are a great source of concern to customers. They go above and beyond their job descriptions to learn about new technologies, and they collaborate with employees in different parts of the world. They practice 3-box thinking by continuing to deliver on their commitments while attacking new problems.

And last, but certainly not least, Denis did what all intrapreneurs aspire to do.

He delivered.

SECTION III: CORPORATE INNOVATION

12. The Corporate Case

Stories of intrapreneurial success at a large corporation must be balanced against the reality of existing corporate processes. Problems, delays, and frustration can occur when a global intrapreneur bumps into a traditional corporate framework.

An experienced intrapreneur shows a great deal of savvy when it comes to avoiding, navigating, and overcoming corporate red tape. They ignore long-held beliefs that innovation is centered in corporate headquarters and take their own initiative to innovate locally.

Corporations that do not find, stimulate, nurture, and enable their global intrapreneurs are greatly limiting their ability to innovate. Products built for global markets will fall flat without a rich network of worldwide intrapreneurs. It is impossible to build environmentally friendly, locally compliant, and globally sustainable products without leveraging the network.

Consider again the five phases of reverse innovation suggested by Vijay Govindarajan, as depicted in Figure 62.

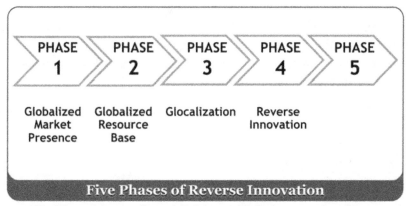

Figure 62. Five phases of establishing a global network of innovation.

A large number of global companies are currently navigating these phases. Many have already reached Phase 3. They have successfully built an international network of developers who have learned to successfully build and deliver products. The genesis of these products typically occurs at corporate headquarters; the products are then glocalized (modified for consumption in local markets) overseas.

Reverse innovation occurs when the corporation moves beyond glocalization. It is not easy for headquarters to cede innovative control. It does not just happen. It requires an intentional modification of the corporate culture. This change, of course, must occur gradually.

The biggest obstacle for moving beyond Phase 3 is the command-and-control methodology that is the legacy of many large companies.

Author Gary Hamel, in his book *The Future of Management*, provides a framework for changing a command-and-control culture to an environment of empowered innovation. His ideas and insights can be applied to the problem of transforming a mindset of glocalization to a corporation that executes on the vision of reverse innovation.

Gary Hamel's bottom line is that corporations need to innovate, but not only in the "new product" sense. They must innovate in their methods of managing employees! In the intrapreneurial context, this means that corporations must manage their intrapreneurs using methods and processes that have not yet been created.

Another way of saying this is that intrapreneurs can innovate all they want, but the true innovative potential of global intrapreneurs can only be unlocked by innovating at the management level. Consider the innovation stack depicted in Figure 63.

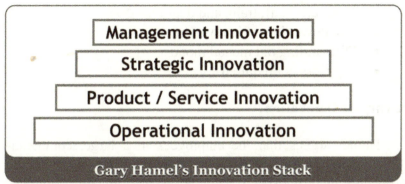

Figure 63. Management innovation has the highest priority for enabling overall corporate innovation.
Reproduced from Gary Hamel, *The Future of Management* (Boston, MA: Harvard Business Press, 2007), 32. © Gary Hamel.

Hamel claims that operational innovation rarely leads to a competitive advantage. He also states that product/service innovation will rarely "grant a company long-lasting industry leadership" (Hamel, *The Future of Management*, 33). Strategic innovation, or "bold new business models" (Ibid., 33), may give a temporary competitive advantage that is easily decoded by others.

The most effective form of corporate innovation, according to Hamel, is management innovation. Management innovation is defined as anything that "substantially alters the way in which work is carried out, or significantly modifies customary organizational forms, and, by so doing, advances organizational goals" (Hamel, *The Future of Management*, 19).

If the goal is to enable and sustain global innovation by harnessing the efforts of distributed intrapreneurs, then it will require altered work habits and/or new organizational forms.

The following list contains more of Hamel's thoughts on management innovation. It is worthwhile to analyze the application of each topic to highlight the need for a new method of managing global intrapreneurs.

- Efficiency or innovation?

- Motivate, staff, train, and deploy.
- The power law of innovation.
- Dismantling old mental models.
- Big risks not required.

Efficiency or Innovation?

Hamel argues that the employees in a large, command-and-control corporation are "captives of a paradigm that places the pursuit of efficiency against every other goal" (*The Future of Management*, 12). Innovation takes a back seat to efficiency.

If this is true, then the traditional goal of a global corporation is to make sure that all global locations are maximizing their productivity. In other words, global employees should be performing Box 1 activities. They should focus nearly 100% of their time developing and delivering products that are currently generating revenue for the corporation.

But as we've already discussed, corporate intrapreneurs refuse to stay in Box 1. They take their own initiative to distribute their time in Box 2 and Box 3 as well. The tales of intrapreneurs reported here demonstrate how these folks repeatedly depart Box 1 to the benefit of their corporations. They have proven that they can both deliver and innovate.

Corporations, therefore, need to hold up the intrapreneur as a local example of a new way to operate. They must herald having found someone who can be efficient and innovate at the same time. In fact, their behavior dovetails quite nicely with Hamel's next point about replicating this mindset to the rest of the organization.

Motivate, Staff, Train, and Deploy

Hamel posits that long-term military advantages were realized by armies that "were able to break with the past and imagine new ways of motivating, staffing, training, and deploying warriors" (*The Future of Management*, 25). He calls these armies "management innovators"

(Ibid., 25), and goes on to give several examples of armies that sustained many years of advantage due to changing their motivational methods.

Global corporations can easily identify the intrapreneurs in their midst. These individuals are productive, well-known within the trenches, and approachable. It is a straightforward affair to ask them to transfer their enthusiasm by being a mentor to others.

Intrapreneurs, more than anyone else, love to see their ideas come to fruition. They are intrinsically motivated by the process. They also recognize that they must depend on others to help them see their idea through to the end. They are willing, therefore, to vocalize their seven highly effective steps as a way of training their co-workers to practice the same behavior. Corporations that encourage their intrapreneurs to become mentors in this way are essentially granting employees the right to spend some of their time outside of Box 1.

When more and more employees spend more and more time exploring outside of Box 1, the corporation can begin to leverage what Hamel calls the power law of innovation.

The Power Law of Innovation

Every company needs a rich portfolio of options for future product and market opportunities. Intrapreneurs provide these options, but not at the level of scale that is required. Intrapreneurs across the globe may generate dozens of ideas; employees worldwide could potentially generate thousands.

Hamel describes it this way:

Innovation follows a power law: for every 1,000 oddball ideas, only 100 will be worth experimenting with; out of those, no more than 10 will merit a significant investment, and only two or three will ultimately produce a bonanza. (*The Future of Management*, 45)

Most companies don't have the ability or the processes to generate this type of portfolio. Yet Hamel argues this is exactly what is practiced by venture capitalists. They sift through "thousands of business plans, meet with hundreds of would-be entrepreneurs, invest in a dozen or so companies, and then hope that one or two of them will become the next Google, Cisco, or Amgen" (Ibid., 45).

Shouldn't companies leverage their intrapreneurs in the same way? If intrapreneurs are walking examples of great ideas and great implementations, shouldn't they be given the edict to inspire others to do the same? If they are allowed to do so, they will, in Hamel's words, "build a diverse portfolio of nonincremental strategic options" (Ibid., 46).

The problem, of course, is that intrapreneurs who behave this way might bump into a corporate reality that could stop them cold: line managers. Line managers represent efficiency in an organization, and they do not necessarily want their employees to leave Box 1 for any reason whatsoever. They are likely to continue functioning in what Hamel calls the old mental model.

Dismantling Old Mental Models

Hamel claims that "Innovators are, by nature, contrarians" (Ibid., 53). The same thought, therefore, applies to intrapreneurs.

Nobody can destroy the old mental model of Box 1 conformity better than an intrapreneur. Intrapreneurs should lead the charge for change, and their corporations should let them get away with it.

The global employee base needs one of its own to step forward and advocate for change. Without this change, the employee base at large will not feel empowered to spend the time to contribute their valuable intellectual capital to improving the status quo. Hamel argues that corporate founders (who were once contrarians themselves) are inclined to resist contemporary contrarians. When this message of resistance trickles down to the employee base, it has negative consequences, according to Hamel:

> It's hard for founders to credit ideas that threaten the foundations of the business models they invented. Understanding this, employees lower down self-edit their ideas,

knowing that anything too far adrift from conventional thinking won't win support from the top. (*The Future of Management*, 54).

The overall goal of shaking things up is to transform a centralized corporation into a globally distributed innovation machine led technically by a united set of distributed intrapreneurs. On one side is the old guard: executives and line managers who staunchly protect their existing processes. On the other side sit the contrarians: the intrapreneurs who can inspire and guide the masses into new markets. In the middle sit thousands of employees who believe that the old guard will win.

It is at this point that Hamel offers his most sage advice.

Big Risk Not Required

Hamel posits that the stalemate that often occurs when trying to guide a big ship in a new direction does not necessarily imply the game is over. The selling of a new innovative model is actually quite easy because it offers the employee base a new lease on life: three new areas of opportunity for them. These opportunities can be found in Figure 64.

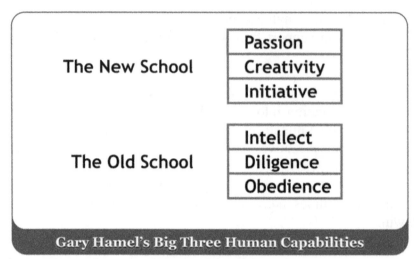

Figure 64. Passion, creativity, and initiative are new values that can encourage global innovation in the old-school employee base.

Hamel argues that today's corporations have squeezed every possible ounce of efficiency from their employee base. As a result, the most valued employee traits are obedience, diligence, and intellect. Taken together, another word emerges from these traits: bored.

The recommendation, then, is that organizations begin to value new traits in employees: passion, creativity, and initiative. In fact, these three traits are perfectly and repeatedly demonstrated by an effective intrapreneur. Intrapreneurs take the initiative to be creative, and they use their passion to drive their ideas to a deliverable product.

While Hamel concedes that the problem is huge, he also states that a corporation doesn't necessarily have to take big risks to solve big problems. As he put it, "if the problem is big enough, progress of any sort will be considered valuable" (*The Future of Management*, 38).

Passion, creativity, and initiative, taken together, result in another word: fun. Aversion to that simple little word sums up why large companies have such a hard time transforming themselves from models of efficiency into agile models of innovation.

The Corporate Case

In its journey to transform itself into a globally innovative powerhouse, EMC took its first step by having fun.

The company sponsored a global science fair.

13. Transformers

By the end of 2006, EMC had one very good problem and one very bad problem on their hands.

On a decidedly positive note, its leaders had successfully navigated the dot com bust earlier in the decade and emerged with flying colors. Sales of its information storage devices were soaring. EMC had the wisdom to diversify its technology portfolio by acquiring a set of technologies that complemented its information storage sweet spot.

The sour note was related to the positive one. In 2006, EMC had acquired nine different companies, bringing the total to 23 companies over the previous 4 years. The number of business units within EMC was soaring. Most of them operated in their own silos. The employees generating the bulk of EMC's revenue stream (information storage devices) were consumed with their own deadlines and roadmaps. EMC was rapidly expanding its R&D footprint outside of the United States.

Local intrapreneurs in the United States were spearheading innovation in their own business units. Several were scattered throughout the world (mainly at foreign companies that had been acquired by EMC).

EMC's Office of the CTO surveyed this landscape and recognized that the company was not positioned for breakthrough innovation on a global scale. It was difficult if not impossible to pry away the intrapreneurs from their business units and allow them to pursue global collaborations in worldwide markets. EMC's historic ability to get the job done at all costs was the epitome of command-and-control efficiency. Allowing employees in the trenches to work outside Box 1 would take away from that efficiency.

Transformers

In a wonderful example of Gary Hamel's motto, the Office of the CTO recognized that any progress on such a big problem is worth it.

So they decided to announce a worldwide innovation contest. The contest was meant to accomplish several purposes:

- It would give employees permission and a level playing field to enter an idea from any corner of the globe.
- It would provide the corporation with a portfolio of new ideas and directions that the company could choose to fund (or not).
- It would serve as the company's first worldwide technological gathering since the company began opening global COEs.
- It would broadcast an important message to all employees: You need to innovate, and you have the corporation's permission.

Every show needs a master of ceremonies and the first step would be to find the right person. The company turned to an employee from a one of its 2006 acquisitions: Dr. Burt Kaliski. Kaliski was an employee of RSA. While RSA certainly knew how to deliver complex software, it also had a culture of innovation and collaboration. RSA had strong university collaborations, as well as a disciplined approach to documenting their new ideas and presenting them as research papers at international conferences.

Dr. Kaliski formed a team of conference organizers and the contest was announced in the spring of 2007. The team outlined a "call for ideas" and displayed the carrot: 30 finalists would travel to the corporate headquarters in October for a "science fair" competition.

Reaction to the announcement was mixed. Most of the 30,000+ employees declined the opportunity to participate. Many of them rolled their eyes and went back to work. It's fair to say there was a good deal of skepticism across the tried-and-true employees in the trenches in the United States.

Overseas, however, it was a different story. Employees not only viewed the contest as a potential great honor within their facility, but also jumped at the chance to win a free plane ticket to the United States. The fact that any employee could participate began to stimulate creative thoughts throughout the global employee base.

Over the next several months, employees in China sent in 18 ideas. Those in Ireland and Israel contributed 24 apiece. Employees in India sent in 62. Ideas flowed in from all corners of the world, including Russia, Australia, Argentina, New Zealand, Germany, Belgium, Singapore, Canada, Japan, and the United Kingdom.

The enthusiasm of overseas idea generators eventually spurred the employees in the United States into action. By the end of the summer, they had contributed a total of 205 ideas. Figure 65 highlights the contest's geographic span.

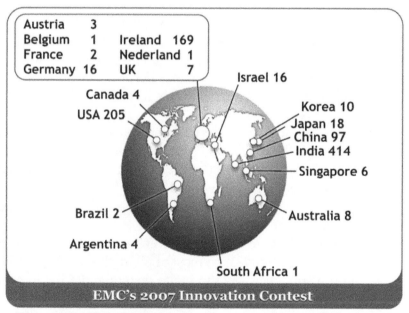

Figure 65. Global involvement in an employee idea contest.

By the end of the summer of 2007, the 30 finalists had been selected. Dr. Kaliski and team decided to take the new culture of innovation to the next step by planning the following 2-day agenda:

- An overview of corporate strategy by the company's top technologists.
- A keynote speech by Don Tapscott, author of the book *Wikinomics*.
- A personal introduction (for the first time) of EMC's global COEs.
- Guest university lecturers from Carnegie Mellon, Indiana University, and UMASS Amherst.
- A presentation from one of EMC's top customers, Credit Suisse.

The first-time event was buoyed by a new and different kind of energy within EMC. Global innovators stood side by side to present their ideas to corporate executives. Busy employees disengaged from their business units for 2 full days to consider new directions for their products. Global employees dialed in to listen to the speakers and support their local representatives.

Perhaps the most strategic decision made by the conference organizing team was the invitation list to the conference. In addition to the 30 finalists, EMC invited 400 individuals to attend the event in person. Each business unit was awarded a specific number of seats to the conference, and the leaders of these business units were instructed to invite a particular type of employee.

They were to invite employees who were technically strong, highly collaborative, inventive, and most of all, highly productive.

In other words, they were to invite the corporate intrapreneurs. For the first time in the history of the company, corporate intrapreneurs were gathering together for one purpose: to discuss innovation.

When the conference was over, the winning ideas were "funded" by assigning a team of developers in the Shanghai COE. These developers worked closely with the inventors and delivered a demonstration to EMC customers at EMC World, the annual gathering of EMC customers and EMC engineers. Global developers collaborated on new ideas and ran them by customers.

This, in essence, is the model for global intrapreneurship. The Office of the CTO chose a contest method to communicate a template for global innovation. This model was not wholeheartedly embraced overnight, and that was part of the reason why the company repeated the contest in 2008 and again in 2009.

By 2009, the contest's profile had achieved visibility to the point that Bangalore, India, hosted the global event. Even more telling of the model's acceptance, multiple sites around the world hosted their own "mini-summits" by inviting in local speakers and customers to spur innovation. Videos from all locations were posted on EMC's central social media backbone for easy viewing by all employees.

The 2009 version featured more extensive employee involvement. Finalists recorded themselves pitching their ideas, and published these pitches globally. Volunteers organized local events in Beijing, Shanghai, Tel Aviv, St Petersburg (Russia), Rotterdam, Paris, Cork, Montreal, and six different locations in the United States. Over 1,400 ideas were submitted by global innovators in 19 different countries. The entire EMC workforce was invited to comment on any and every idea. For the first time, employees were allowed to vote for the top 100 semifinalists. Figure 66 shows the results of the 2009 contest.

Figure 66. The 2009 content witnessed increased global involvement in corporate innovation.

In just 3 years, EMC had made great strides in transforming the innovative mindset of its global employee base. The result depicted in Figure 66 meets Gary Hamel's idea of the power law of innovation. The corporation as a whole garnered a portfolio of 1,422 ideas to peruse. One hundred of them were promoted by global employees. Thirty of them were promoted by a hand-picked set of technology experts. Ultimately, six of them were chosen for direct funding (once again using international cooperation to advance the idea).

This result, however, still falls short. In an organization that had grown to over 40,000 employees, with hundreds of different products, and millions of different customers, 1,400 ideas are only a tiny trickle in the pipeline. Somehow, the building of a portfolio of innovation needs to be extended throughout the organization. One department (e.g., the Office of the CTO) cannot scale to handle this type of effort.

This problem is best solved by applying the template created at the corporate level and implementing it within the local business unit. In 2009, EMC intrapreneurs in global locations began to assume responsibility for administering innovation contests within their local business units. The requests to administer these contests, of course, are met with varying degrees of resistance. These objections are overcome by pointing to the corporate example and, when necessary, explicitly stating the obvious: "The company is asking us to do this."

In response to the call, an intrapreneur in Shanghai held an innovation contest for the entire facility. Ideas ranged from new product suggestions, to better processes, to environmental sustainability. The employees in Shanghai wrote contest software to make it easier to submit, view, and comment on entries by their peers (this contest software was eventually adopted as the official 2009 corporate contest software) – an example of innovation supporting further innovation.

A global coding challenge was run by an intrapreneur in the employee's 2,000-person business unit. One winner was flown in from the United Kingdom and another was flown in from St. Petersburg, Russia. They both received awards in front of their technical peers at the annual innovation showcase.

An intrapreneur in the United States ran a business unit contest for employees in four different locations. This contest was held at the worst possible time: during the stretch run of delivering a product to customers. Despite the pull to complete the deliverable, employee participation was strong, with the majority of ideas again being generated overseas (in St. Petersburg).

In general, business unit contests tend to follow a very specific pattern as far as categorization of ideas is concerned. Figure 67 depicts how idea submissions fall into the 3-box model's categories.

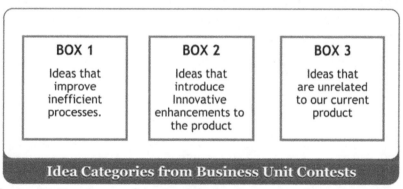

Figure 67. Employees from business units tend to focus strongly on Box 1 suggestions.

The rationale for running a contest during the most stressful and inopportune time (the end game of delivering a product) is that the development team is at its most frustrated due to inefficient business processes and lack of resources. Idea contests can serve as a repository for improvements that need to be made to eliminate internal bottlenecks. In fact, it has been suggested within EMC that contests be made a critical part of the development cycle!

The bottom line is that different business units now have their own portfolio of ideas that can be pushed forward into the future (again, Hamel's power law of innovation in action). As intrapreneurs across more business units begin to take the responsibility for stimulating innovation within their own peer groups, scalability is achieved and the company has a wealth of idea options from which to choose.

With one small step, EMC set into motion a new culture of innovation. A dedicated team identified global intrapreneurs, brought them together, and instilled a vision of a new way to innovate globally. This approach works because it relies on the efforts of individuals who are already widely respected within their organizations.

Innovation contests, however, were only the starting point. Momentum ultimately fails if global intrapreneurs' ideas do not result in investments in continued global collaboration. The next part of the strategy involves getting to know what makes an intrapreneur tick.

The corporation needs to know how to catch them and reel them in to foster even more enthusiastic and productive intrapreneurs.

14. Reel Them In

Intrapreneurs are born contrarians. They also tend to avoid widespread corporate visibility, in part because it's a distraction from their ultimate objective of shepherding an idea through to delivery.

Given these realities, corporate engagement of global intrapreneurs becomes a difficult dance to choreograph. By nature, intrapreneurs can be skeptical of corporate directions and plans because they often see a different vision from the perspective of their trenches.

Given the importance of engaging intrapreneurs, what is the most appropriate strategy to "reel 'em in" like a fish on a line?

The best way to motivate intrapreneurs is to present a win-win scenario via corporate give-and-take. The corporation gives ever-increasing support to achieve an intrapreneur's primary objective: the delivery of their ideas. The intrapreneur, in return, agrees to engage with the executive level on the innovative health and future of the company.

The first step in this program is to formally recognize and develop global intrapreneurs. At EMC, this step has already been taken with the creation of the Distinguished Engineers and Fellows program. Intrapreneurs who function at a high level within their business units are first nominated for selection as a distinguished engineer (DE). Once they are recognized, a bargain needs to be struck.

The initial conversation between new inductees and a corporate executive should sound something like this:

Congratulations and thank you for agreeing to come to this meeting. I know that your time is valuable, so I'll be brief. This is the last meeting where I will mandate your presence.

You will have the option to access the resources in my office. You are hereby authorized to make frequent customer visits at your discretion. You are hereby authorized to visit your global colleagues overseas at least once per year.

You will be granted 12 months' worth of co-ops or student interns to advance your ideas. In return, I expect you to freely and openly provide access to all of your ideas and all of your work. I expect you to provide tangible mentorship to new employees. I expect you to assist in corporate recruiting of top technical talent. I expect you to engage in university research and present at least one paper per year. I expect you to collaborate with each other on your own.

Finally, from now on, you need to invite me to meetings because I need to know the technical direction in which we should take this company. You own the roadmap.

Let's break down this little speech into its give-and-take components. First, it is important to analyze what is being given to the intrapreneur.

Freedom

The intrapreneur, having proven he or she intuitively knows the right thing to do, is formally given the freedom to do the right thing without further encumbrances. The intrapreneur is not mandated to participate in a never-ending array of bureaucratic meetings. He or she is not expected to attend late-night corporate strategy sessions in lieu of decompressing, thinking, or recharging personal batteries.

Opportunity

Corporate training, industry experts, corporate councils, technology and strategy sessions, corporate speakers, industry events, and university lectures are corporate resources that should be offered regularly to every intrapreneur. Given the foundational right of freedom that has been bestowed upon intrapreneurs, these resources are optional and opportunities that are there for the taking.

Customer Visits

The seed of any intrapreneur's strategy is often the desire to solve a problem presented by a customer. Perhaps the most critical entitlement of an intrapreneur is the right to engage directly with customers whenever and wherever possible. In addition, intrapreneurs should have unfettered access to any request for enhancement (RFE) lists submitted by customers.

EMC has defined the title of a DE to include those field engineers who are outstanding advocates for customers. These field engineers are intrapreneurs in their own right; they experience continued exposure to customer problems as well as a wide array of adjacent technologies. For this reason, the corporation "expects" that developers and field engineers will self-organize at the corporation's various local, regional, and global events.

Face-to-Face Global Collaboration

Telepresence is a wonderful thing. It allows intrapreneurs to visually interact with their global colleagues in real time. However, it falls short in its ability to form strong (and in some cases lifetime) bonds between innovative collaborators. When an expert in one technology makes a visit to a foreign location, that expert's time is exponentially valuable. The ability to address large numbers of foreign employees in person allows for the dynamic and free-flowing exchange of ideas that characterize innovation.

Furthermore, global travel is a reward for those people who labor in the trenches of product delivery. The frequent model of a visiting intrapreneur inspires local employees to rise to the same level of idea-to-deliverable achievement.

Global travel to remote offices is a pricey corporate advantage that cannot be replicated at a startup! Intrapreneurs should be able to leverage this advantage.

Interns and Co-ops

The offer of staff hours as a benefit hits the intrapreneur right in the sweet spot; having the ability to hand-pick and hire assistants to carry out ideas, rather than inherit someone else's choices of individuals who may or may not have good ideas to contribute, is invaluable. These assistants are not influenced by existing corporate processes and are hungry to build something new. Intrapreneurs are experts at carving out a portion of their time to advance their ideas; corporations can assist by providing resources that spend 100% of their time on the ideas.

These five benefits address and support many of the habits of highly effective intrapreneurs. They represent a strong commitment to the individual. In return for this comment to the intrapreneur, the corporation expects to gain something as well (the corporation "gave," so it reasonable to also "take"). Interestingly enough, each of these takes is mutually beneficial.

Accessibility

Intrapreneurs, whenever possible, should advertise their ideas, prototypes, and thoughts. If the corporation is going to bestow freedom, provide opportunities, and fund customer interaction, travel, and interns, the least the intrapreneurs can do in return is explain what they've experienced.

There is no better forum for this return than social media tools like blogging, wikis, Twitter, and so on. It's a simple request! The benefit will pay dividends to the intrapreneur; adjacent technologists can interact and enhance the directions being advertised.

Mentorship

The corporation wants to create not only a pipeline of ideas, but also a posse of intrapreneurs to generate those ideas. There is no better way to do this than to rely on the transfer of innovative behavior and thought processes from experienced employees to junior employees.

Mentorship can benefit the intrapreneur who chooses to leverage the experience in a unique way. In many cases, the intrapreneur can "claim" a new employee for the first 3 months of the new hire's career. The intrapreneur can ask the new employee to advance the cause of a new idea before turning him or her over to the specific role for which the new employee was originally hired. This exposure to innovation provides lasting benefit to the new hire.

Corporate Recruiting

Intrapreneurs are typically not rock stars. In many corporations, they are relatively unknown because they prefer to stick to the trenches. However, if they have a track record of innovation, heads will turn when they speak in front of college students and potential new hires. Hiring the top technical talent is a huge win for a large corporation.

In the course of participating in recruiting, the intrapreneur may become a frequent guest at local college campuses. Involvement with the local college has two benefits:

- The intrapreneur, and hence the corporation, becomes familiar with the academic research on the campus.
- The intrapreneur, and hence the corporation, becomes familiar with some of the top students on the campus.

The intrapreneur who finds areas of overlap with local universities then has yet another avenue to progress novel ideas and research. Recruiting, even if it's somewhat informal, goes hand in hand with the corporate desire for intrapreneurs to conduct university research.

University Research

For every new idea discussed inside the walls of a corporation, there are likely one or more academic initiatives that are already tackling the same (or a similar) problem. When an intrapreneur becomes exposed to these initiatives, an important dynamic takes place: idea acceleration.

As intrapreneurs around the globe familiarize themselves with a topic being researched, they advertise it by making it available to the global employee base at large. This advertisement is consistent with their agreement to advertise all of the work that they do.

The chance to visit college campuses, both local and remote, is a reward. At EMC, the corporation is taking action to formally build virtual teams with universities. Figure 68 depicts the steps taken by these virtual teams to build up strong university relationships.

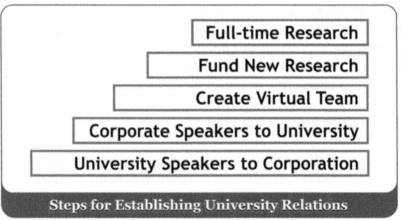

Figure 68. Steps in a startup model to involve intrapreneurs with local university research.

This model begins by having a corporate intrapreneur establish contact with university professors. These professors are then invited into the corporation for a presentation/lecture on their specific area of research. The intrapreneur, leveraging connections with employees in the trenches, issues a general invitation to the organization (and specifically invites the employees who would have a strong interest).

These local visits from professors are then followed by on-campus lectures given by local employees. The two-way transfer of knowledge between corporations and universities stimulates a potential set of innovative new research possibilities. Those employees who wish to pursue the research further can ask to be part of a virtual team.

Virtual team members are specifically nominated by the head of their business unit (e.g., a senior vice president). The corporation, by allowing employees to participate in university research, is explicitly allowing their employees to venture outside Box 1.

The virtual teams, over the course of time, will likely generate mutually beneficial work proposals with their academic counterparts. These proposals can turn into requests for funding by the universities, and an intrapreneur takes on a new role akin to that of an entrepreneur finding funding for a startup. Intrapreneurs, however, typically need much less funding than a startup company to progress these new ideas.

The final stage of university relations occurs when funded research begins to yield (or promise) revenue results for the corporation. If the intrapreneur has been successful in stimulating research that benefits his or her business unit's own product lines, the case can be made to fully fund university research through the hiring of a dedicated university researcher. This researcher will assume a full-time role in advancing corporate causes in a university setting.

At each of these five stages of the model, the intrapreneur is advertising all of the lectures, the ideas, and the funded projects. This information is made globally accessible within the corporation. Geographically distributed intrapreneurs become aware of these different research venues.

The win-win scenarios presented in this chapter can be summed up by using one word: progress. The ease with which an intrapreneur can make progress is a self-motivational corporate carrot that is dangled in front of an intrapreneur. The corporation provides intentional avenues for supporting the ideas of the intrapreneur, and in turn, the intrapreneur pushes his or her innovation agenda and begins to inspire local co-workers.

The intrapreneur, with the blessing of the corporation, is officially recognized as what Daniel Pink calls a Type-I employee. This is Pink's description of Type-I behavior: "Type I behavior is fueled more by intrinsic desires than extrinsic ones. It concerns itself less with the external rewards to which an activity leads and more with the inherent satisfaction of the activity itself" (*Drive*, 63).

The next corporate step is to produce more Type-I employees. Type-I employees are motivated by passionately working on interesting tasks that they can call their own.

Pink asserts that Type-I employees are made and not born. Can corporations bottle and distribute the characteristics of their corporate intrapreneurs?

15. Growing the Type-I Population

Consider a corporation that has identified its existing intrapreneurs and recognized them.

The job is only half done, and the easy part is over.

Uniting intrapreneurs globally is a logical (but difficult) next step. Strategies for accomplishing this will be covered in the next chapter.

An equally challenging task is the continual development and growth of worldwide intrapreneurial talent. Existing intrapreneurs can serve as role models and mentors, but this sort of diplomacy is not enough. Corporations need well-establish methods that provide a steady stream of new blood.

This chapter suggests a corporate framework for intrapreneurial development based on the seven habits of intrapreneurs. Each of these habits can be taught to any productive employee who shows a glimmer of innovative potential.

Listed here are the seven habits of highly effective intrapreneurs:

1. Productivity
2. Initiative
3. Collaboration
4. Awareness of the 3 boxes
5. Navigating visibility
6. Bridge building
7. Finishing

Daniel Pink's book, *Drive*, lays out the reasons why the roll-out of intrapreneurial values is so difficult to achieve. Large corporations have trained their employees, above all else, to be efficient. They value algorithmic thinkers who can repeat tasks at increasing levels of speed. They use what Pink calls an "if. . . then" approach to the completion of tasks. This technique uses extrinsic motivation. If you complete the task, then you will be rewarded (e.g., with money, promotion, recognition).

In other words, most existing corporations are already training their employees to practice the foundational habit of an intrapreneur: productivity.

Habit #1: Productivity

Corporations do not want to stop training their employees in how to be productive. What they do want to avoid, however, is solely focusing on extrinsic rewards in a never-ending quest for employee efficiency. Pink describes the logical conclusion to the continued use of extrinsic awards: "For some people, work remains routine, unchallenging, and directed by others" (*Drive*, 29).

Pink goes on to state that this type of work can also play on employee fears: "Routine work can be outsourced or automated; artistic, empathetic, nonroutine work generally cannot" (Ibid., 16).

Many productive employees often look at corporate intrapreneurs and wonder how they always seem to be working on "the cool stuff." It is true that young employees grow to be experts by completing "algorithmic tasks at the beginning of their careers" (Ibid., 46). These tasks are often best rewarded by carrots such as monetary raises and/or career promotions.

Corporations, however, need to make a fundamental shift once an employee has been recognized as highly productive. The first step in the making of a Type-I employee is to provide them with the opportunity to become more autonomous. According to Pink, Type-I autonomous employees work on their own terms. Indeed, Pink describes autonomy as one of the key nutrients that leads to Type-I behavior.

Corporate intrapreneurs may not know they're already practicing autonomy because they call it initiative.

Habit #2: Initiative

Initiative is the key attribute that separates intrapreneurs from highly productive employees. Recall that intrapreneurs apply their initiative to two focused areas: customers and adjacent technologies. Figure 69 depicts again the first opportunity for shaping productive employees into Type-I practitioners.

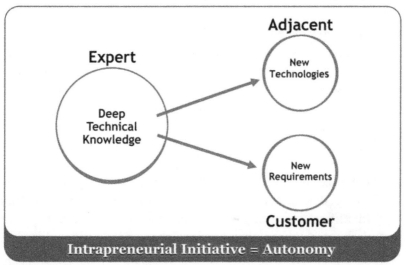

Figure 69. Productive employees can be offered the opportunity to become autonomous.

Productive employees have typically finished their assigned work and delivered a desired product or service. In most large corporations, they are either refocused on the next product version or reassigned (or transferred) to a completely new project. Whatever the case may be, the corporation should leverage this critical juncture of their career by assigning them Type-I goals:

- Go visit a customer.
- Learn something new.

If either or both of these goals are added to an employee's goal sheet, they should be added just as they are written here. Don't tell them which customer to visit, and don't tell them what to learn.

At this point, productive employees should have their first collaborative experience on their Type-I journey. They should speak to their local intrapreneur.

Habit #3: Collaboration

Intrapreneurs, when they first mentor productive employees, are not project managers handing out tasks. They shouldn't have ready-made projects that are given out like a pablum, the same task to every individual. Instead, they offer a survey of the corporate landscape.

They talk about customers that they have visited lately.

They talk about industry bloggers that they are following.

They talk about research being conducted at the local university.

They talk about recent collaboration with their global peers.

During this conversation, the intrapreneur assists the mentee with the identification of adjacent fields in which mentees may take an interest. Perhaps most importantly, intrapreneurs identify a list of local customers who can provide insight into new problems that the mentee can help solve.

Most corporations, if they do send young employees into the field, will send the employee to a customer who is using the employee's recently built product. There is nothing wrong with this approach.

For the development of Type-I employees, however, it is best to visit customers who are using (or need) different products than those with which the employee is familiar. This exposure to new problems and new technologies leaves the employee with a set of questions that cannot be answered independent of collaboration. They often have a partial answer based on their level of expertise, but they need to perform an adjacent search to generate a fuller solution. The idea is to nudge the productive employee to take initiative to investigate the appropriate adjacent technology.

This search, once again, can be guided by an intrapreneur. Intrapreneurs can steer the employee towards training and/or certification. They can recommend visits to local professors. They can provide references to academic papers and/or industry white papers.

Most importantly, mentoring intrapreneurs can provide references and set up meetings with their many contacts throughout the corporation.

The end result of these guided collaborations is an employee who has gained experience in working with others to generate a new idea that solves a real customer problem. These employees have satisfied the goal of visiting a customer and learning something new, and they have learned a new skill: self-direction via autonomy.

Keep in mind that this autonomy is not born of independence. As Pink points out, autonomy means *inter*dependence. The employee has learned to rely on intrapreneurs and the contributions of customers and co-workers.

Successful realization of autonomy begets a second characteristic of Type-I employees: purpose. Autonomous employees have their own mission at work that was not born of a dictate handed down from a corporate executive; it was handed down by a customer with a need. Purpose now intrinsically drives these employees to devise a solution to please the customer.

It is at this point that the employee reaches the most critical phase: building the first realization of the idea. Indeed, this phase is littered with would-be intrapreneurs who could not overcome the corporate culture of Box 1 thinking. If every employee were allowed to work solely on their new ideas, the funding for these ideas would completely dry up.

Still, at some point, the employee needs to make a first attempt at 3-box behavior. Intrapreneurs in training must simultaneously contribute to the bottom line while proving that their new idea has merit.

Habit #4: The 3 Boxes

Figure 70 is a reminder of the framework in which potential Type-I employees must learn to operate.

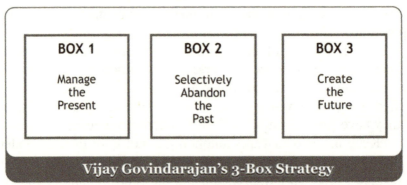

Figure 70. Productive employees must be taught to perform efforts in parallel among the three boxes.

The tension between staying in Box 1 (what your boss wants you to do) and innovating in Box 2 and Box 3 (the employee's new purpose) is the true obstacle for large corporations and it's the stumbling block that most often stops intrapreneurs in training. Ideas often die (and employees become discouraged) at this stage. Employees will soon realize that the pressure to stay in Box 1 will make it very, very difficult to make progress on their own idea.

It is at this point that the involvement of the intrapreneur is most critical for success in development of the next generation of Type I employees. Intrapreneurs have pledged to take responsibility for progressing innovation amongst employees within their business unit, but how can they possibly guarantee that each new idea is given adequate attention?

Intrapreneurs leverage two things to accomplish this goal:

- Plus-2 visibility.
- The power law of innovation.

Intrapreneurs, by necessity, have built a strong relationship with their management chain two levels above them. The top-level manager is typically a director or senior vice president. Often, this manager looks to the intrapreneur to determine the course of future roadmaps. The intrapreneur, at this juncture, can present a portfolio of proposals that are fueled by customer need and employee passion.

A meeting is scheduled. It may occur monthly, or it may occur quarterly. The employee is put on notice: Your idea will be presented to the division's vice president on a certain date.

There's nothing like a deadline to motivate Box 2 and Box 3 behaviors. Notice, however, that the reward is not monetary. The reward is feedback. Someone in a position of authority will listen to the employee's idea and provide feedback. This intrinsically motivates the employee to do their best work. Nobody needs to monitor them. They use their autonomy and purpose.

The next roadblock they will likely run into is their lack of time.

Habit #5: Navigating Visibility

Pink has highlighted autonomy and purpose as two of the primary drivers that form Type-I employees.

The third and final quality of a Type-I employee is mastery. Type-I employees must master their ability to manage their time and simultaneously deliver two types of organizational value:

- They must deliver on Box 1 revenue.
- They must deliver a presentation of their idea to their organization.

Pink points out that Type-I employees don't "want to let down their current teammates by abandoning ongoing projects" (*Drive*, 91). How do they find the time to work on something new when their co-workers are expecting productivity?

This is a make-or-break time in the formation of a Type-I employee. Their manager can make this easy, or their manager can make this very, very difficult.

Pink describes the ideal manager of a Type-I employee as one who provides "autonomy support":

> These bosses saw issues from the employee's point of view, gave meaningful feedback and information, provided ample choice over what to do and how to do it, and encouraged employees to take on new projects. The resulting enhancement in job satisfaction, in turn, led to higher performance on the job. (Ibid., 77)

Higher performance means that Type-I employees can fire on all cylinders and accomplish both of their objectives. What happens, however, when their manager is less Type I and more command and control?

The intrapreneur provides counsel at this point and encourages the employee to take his or her effort into skunkworks (stealth) mode. If the management paradigm of the organization is command and control, then there are definitely bureaucratic and wasteful meetings that can be eliminated from the employee's day. The intrapreneur may share many of the following tricks of the trade:

- Beg out of any meeting called by someone not in your management chain. Send an e-mail with your input.
- Beg out of any meeting called by anyone greater than plus-2 in your management chain. You probably don't need to be there and can find out what happened afterwards.
- Adopt agile techniques with your co-workers. Hold 15-minute meetings in which nobody sits in a chair.
- If you must attend a meeting that promises to be unproductive, dial in remotely, put the phone on mute, and work instead.

- Block out huge amounts of time on your calendar for personal innovation and collaboration. Anyone who tries to schedule a meeting with you that coincides with that blocked-out time will know you're not available.
- Schedule skunkworks meetings in remote conference rooms where you can remain undiscovered and uninterrupted.
- Start your workday much, much earlier than anybody else.

Limiting employee visibility is a direct benefit of Type-I behavior. By mastering these types of time-management techniques, the employee begins to spend more time on their passion and less time chasing extrinsic rewards. The energy spent on innovation brings the employee to the position from which he or she can publicly pitch a self-grown idea for the first time.

Habit #6: Product Bridge Building

Organizations that routinely survey the landscape of employee ideas are clearly practicing Gary Hamel's power law of innovation. Hundreds of ideas result in dozens with potential; dozens with potential will likely result in a small but worthwhile set of real winners.

When productive employees pitch their ideas in front of corporate management, they are practicing the final stage of Type-I development: mastery. If their idea is "funded," they can begin to build the bridge to future product releases within their organizations. If their idea is not funded, they will gain valuable feedback that can help them learn how the idea can be refined.

The employee at this stage has taken the initiative to engage with customers and worked with them to articulate a problem or need. This has given the employee a sense of purpose, and the customer need should be presented clearly during the pitch to executive management. Executive management is keenly aware the need will not go away if the idea is not directly funded or adopted!

Traditional employees may take idea rejection as the final part of their mission and purpose. Type-I employees embrace rejection at part of the process, perhaps even part of the fun. They go back to the drawing board with increased purpose because the path they have travelled to this point is much more enjoyable than the traditional carrot-and-stick approach.

Pink describes this part of the journey as follows: "When the reward is the activity itself – deepening learning, delighting customers, doing one's best – there are no shortcuts" (*Drive*, 37).

Anything worth doing is worth doing right, even in the face of corporate setbacks.

Habit #7: Finish

The final stage of the employee journey to Type-I status brings the newly minted intrapreneur right back to the starting line: be productive. As intrapreneurs in training have traversed the seven habits of a corporate intrapreneur, they have simultaneously learned the three main traits of the Type-I employee:

- Autonomy: They have learned the only way for them to get great work is to take the initiative and go out and find it for themselves.
- Purpose: They work with a sense of mission because they have selected projects, customers, and co-workers who share a passion for solving the same problem.
- Mastery: They have learned how to push their ideas through the corporate machinery.

All that remains is for them to deliver upon their idea (which is no easy task). They already possess the one skill that got them here in the first place: productivity.

Upon finishing the delivery of their idea, they have earned the right to label themselves as an intrapreneur. They have delivered their own idea within a large corporation.

Of course, while they are delivering their idea, they are simultaneously lining up the next one.

Growing the Type-I Population

The growth of Type-I employees can and should be nurtured at all of a corporation's global locations. The final piece of innovating with global influence is to cross geographic boundaries.

The best person for this job is a boundary spanner.

16. Boundary Spanners

This chapter introduces a framework that defines a strategy for uniting global intrapreneurs. Uniting intrapreneurs is not the end goal; distributed, worldwide innovation is the ultimate desired state. Perhaps no corporate role is more suited to implement this framework than a boundary spanner.

A boundary spanner, loosely defined, is someone who has deep expertise in multiple fields, is respected in multiple disciplines, stays in touch with emerging customer problems and technology trends, and often facilitates the handshakes that need to occur between multiple teams in order to get something done (Lynda Aiman-Smith, "Get the Most from Your Boundary Spanners," *NCSU CIMS Technology Management Report,* Winter 2009-2010: 1).

In a global context, a boundary spanner has been defined as "critically important in developing successful strategies in a global, complex, and increasingly chaotic external environment" (Sean Ansett, "Boundary Spanner: The Gatekeeper of Innovation in Partnerships," *Accountability Forum*, 6: 44).

Boundary spanners, like intrapreneurs, bring together adjacent people (e.g., customers) and technologies to solve difficult problems in a mutually agreeable fashion. To enable reverse innovation in a corporate environment, a boundary spanner must solve two main problems:

- Dismantle the "U.S.-only" innovation mentality.
- Stimulate the autonomy of global employees.

Figure 71 depicts the historical corporate legacy of pushing innovation from U.S. headquarters to global locations (using the countries mentioned in this book as an example).

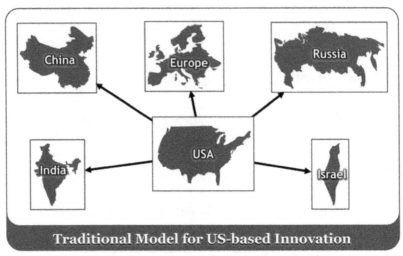

Figure 71. U.S. intrapreneurs push their ideas (and products) to offshore employees for implementation.

The desired new model is depicted in Figure 72: innovation occurs at all sites and product implementations (and sales) can be pushed anywhere.

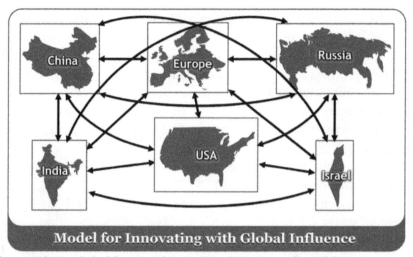

Figure 72. A global innovation network decentralizes idea generation and delivery.

Intrapreneurs, by their very nature, are boundary spanners. They develop an expertise and then take the initiative to acquire more expertise through their collaborations with customers and other technologists. Asking them to become global boundary spanners is a logical extension to their career growth.

Equally important to the success of the boundary spanner is upper-level management at each COE. The management team in place at each COE sees the local big picture of employee capabilities, market presence, and most importantly, the cultural approaches that will best work for their employees (or the ones that will definitely *not* work).

Consider how these two teams (global intrapreneurs and COE leaders) can jointly attack the two barriers to reverse innovation.

Problem #1: Dismantling Centralized Innovation

When a corporation recognizes its need to accelerate employee innovation at all of its worldwide locations, it may fall victim to the centralized mentality.

In other words, smart people at corporate headquarters may design innovative process changes and push the implementation to remote locations. This behavior simply mimics the model that has been used to create these remote locations: Generate an innovative product idea in the United States and push the implementation overseas.

There are many reasons why this approach won't work. The brainpower in the United States cannot fully account for the diversity of cultures across all locations. For example, will Daniel Pink's Type-I behavior research apply in all locations around the world? Or is it more geared towards the western world?

Indeed, the level of experience at each COE – that is, the number of years for which each COE has been in operation –greatly affects the ability of each COE to assume more responsibility for global innovation. Figure 73 depicts the years of experience of six COEs.

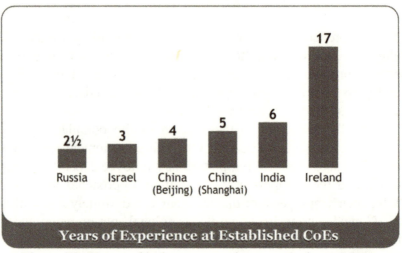

Figure 73. Years of industry experience at various COEs.

In addition to the years of existence, each COE has its own bell curve for employee maturity. Some COEs started up by hiring college graduates with little to no expertise in relevant subject matter. Other COEs were formed by more veteran employees who brought great familiarity with industry trends. Historical hiring practices have a great impact on the ability of each COE to innovate on its own. Figure 74 contrasts the employee experience level between two COEs in terms of career growth relative to engineering titles.

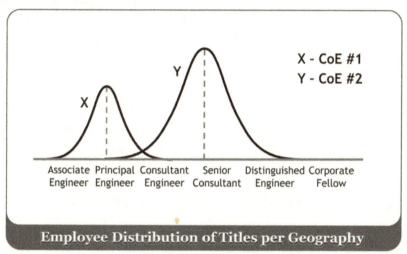

Figure 74. The level of expertise at a COE translates to specific needs for stimulating employee innovation.

These charts and statistics highlight the futility of establishing one single corporate process for global innovation. Does this imply that U.S.-based intrapreneurs should play a passive role in this type of effort?

The answer is no. Corporations employ many intrapreneurs who typically reside in one of two categories: product development and field support. A large percentage of these intrapreneurs are found within the United States, and many are scattered throughout the globe, as highlighted in this book. These individuals may have decades of experience in the local engagement of customers, the germination and incubation of new ideas, and the initiative and persistence to finish the job.

The proper approach, then, for decentralizing innovation is to establish a process whereby the management team from each COE articulates its particular needs to the global intrapreneur community at large. Each COE can then pull the appropriate corporate resource into the COE for the purpose of elevating intrapreneurial expertise.

Some COEs may lack a strong field presence. Some may lack knowledge in a particular technology (or technologies). Others may need intrapreneurial mentoring before they can grow the careers of their most promising employees. The unity of COE management and intrapreneurial talent can find the best fit for each need, but it must be driven on a case-by-case basis for each geography.

The establishment of this new corporate process demands boundary spanners. Their ability to navigate cross-functional teams must naturally extend to the navigation of cross-cultural teams. With their help, each COE possesses critical resources to address Problem #2.

Problem #2: Stimulate the Autonomy of Global Employees

The goal of this process is to empower global employees to effectively innovate in their own markets. The strategy is to leverage global intrapreneurs found elsewhere in the company. Perhaps there is no better way to describe the local implementation of this process than to present a very specific use case: St. Petersburg, Russia. The areas of need highlighted by the Russian COE can be compared and contrasted to those of other locations around the world.

Figure 75 depicts the strategy of "intrapreneurial pull" towards the Russian COE.

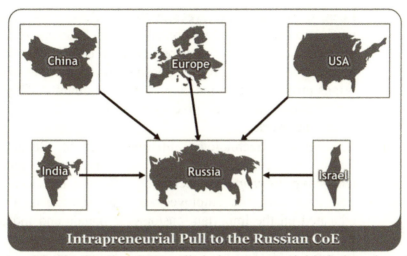

Intrapreneurial Pull to the Russian CoE

Figure 75. The Russian COE pulls in intrapreneurial help based on the needs of its technical population and customer market.

Pavel Egorov is a director of the engineering community at the Russian COE. He has carefully studied the progress and capabilities of his technical team. In May 2010, after 3 years of existence, the St. Petersburg COE had worked on some of the most mission-critical products in the EMC portfolio, including the Symmetrix and CLARiiON disk arrays. The team had already proven it possessed the foundational attribute of intrapreneurs: productivity.

Pavel began to list the requirements for building the COE to the next levels of intrapreneurial development, and has taken steps to match these needs with available talent found elsewhere within the corporation. Some of these needs are as follows:

- University initiatives: The St. Petersburg facility will model their university relationships after other successful locations within the corporation, including EMC Research Cambridge and EMC Research Beijing. Employees have already begun an outreach to local universities, including the St. Petersburg State University, State Polytechnic University, and State University of Information Technologies.

- Local demo lab: Employees at the St. Petersburg facility need to establish strong bonds with the customers (and potential customers) of their products. To do this, Pavel and his team are setting up a demonstration lab that physically brings customers into the engineering community. This lab can replicate the best practices of other customer labs around the world.
- Leading new technology initiatives: The St. Petersburg facility wishes to make the big leap from *building* technology to *proposing* technology. To prepare successful new proposals, the employee base at large needs to be educated on the important customer demands and industry trends in global markets. This education can occur by forming a technology panel and then inviting global intrapreneurs to lecture and collaborate. The strategic directions presented during these panels can be translated into focus areas for the local employee base.
- Career growth: Given the 3-year history of the St. Petersburg facility, the majority of employees have not yet acquired the intrapreneurial habits of global initiative and collaboration. Direct mentoring and sponsorship with successful intrapreneurs can be initiated with any number of global intrapreneurs; these sessions can result in career advice that is the equivalent of a DNA transfer to more junior employees. This advice includes teaching the basics of navigating visibility and intrapreneurial bridge building discussed in previous chapters.
- Idea incubation: Given a new set of technology areas on which to focus and a growing relationship with local universities, the Russia COE can begin the gestation of new ideas made by new employees and/or student interns. This research can be reviewed by global intrapreneurs, which in turn can facilitate global collaboration on related efforts in other geographies.

The initiative being taken by Pavel and the Russian team is exactly what is required to stimulate global innovation. The proposals are made locally and specific help is pulled from the global pool as needed.

Most notably, a global innovation mandate is not being pushed from corporate headquarters.

This framework relies on geographic initiative.

And initiative ranks second only to productivity as a key attribute of an intrapreneur.

17. Egyptian Epilogue

The capacity for technological innovation now extends to the global workforce.

Companies that learn how to globally collaborate are the ones that will win.

There has never been a more exciting time to pursue a career in technology. Whether fresh out of college or a seasoned engineer, innovation is no longer limited to a privileged few. It is no longer limited to one geographic location, and it is no longer best found at a startup.

The technical communities in Beijing, Shanghai, India, Israel, Russia, and Ireland are innovating, along with dozens of other communities in Europe, South America, Asia, Australia, and North America.

Corporations are opening new locations at a dizzying rate to address the emerging markets found in developing countries.

This book, written by a software engineer, is largely biased towards global innovation at the R&D level. Left unaddressed, to some degree, has been innovation by the most critical employees in any corporation: customer-facing employees. These individuals are the ones who talk to customers every single day and deserve a chapter of their own. The customers with whom they interact are the ones who plant the seeds of innovation for any company. Consider again the components of intrapreneurial initiative shown in Figure 76.

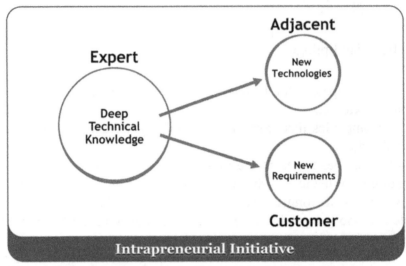

Figure 76. An intrapreneur builds on expertise by learning new technologies and customer requirements.

No matter how much customer exposure an R&D engineer receives, it will never match the exposure experienced by customer-facing employees. It is critical for a corporation to have a strategy that effectively involves field personnel in the intrapreneurial process. Nobody can articulate customer needs better than they can. If they can be trained to innovate globally based on customer requirements, the sky's the limit.

The challenges experienced by field personnel in terms of the breadth and depth of global customer needs are enormous. Consider the long list of issues experienced by the field when trying to cover just three adjacent continents: Europe, Asia, and Africa.

- There are 115 countries on these three continents.
- The total population exceeds 2 billion people.
- There are over 2,000 distinct languages.
- The distance from east to west is 5,793 miles (9,323 kilometers).
- There are 14 different time zones in this east-to-west span.

Egyptian Epilogue

- The cultural norms (not all work weeks start on Monday!) and customer expectations vary widely by region.
- Certification and governmental regulations vary by country.

Field support personnel working at the COE locations discussed in this book have the ability to cover some of the geography within their region, but their reach (and their language skill set) only extends so far.

As a direct response to the customer needs listed above, EMC decided to open a new COE in 2009: Cairo, Egypt. For nearly the last 20 years, two innovative government agencies (the Information Technology Institute and the Information Technology Industry Development Association) have been investing in the building of a skilled IT workforce. As a result, in the last 10 years, Egypt has experienced a stunning transformation:

- From 1999 to 2009, the number of IT companies grew from 266 to 3,032.
- The number of IT trainees grew in the same period from 500 to over 36,000.
- During this same period, the number of Internet users in Egypt grew from about 300,000 to over 13 million.
- Foreign direct investment (FDI) has increased from $2.1 billion in 2003/2004 to a peak of $13.2 billion in 2007/2008.
- The Egyptian workforce is customer-oriented by nature; an enormous amount of Egypt's GDP comes from the tourism sector.

As of this writing, the Cairo COE is staffed with multilingual, product-savvy field engineers. This COE is expected to quickly become a pivotal epicenter for the collection of customer conversations. Employees in these locations must be transformed into intrapreneurs; they must begin to transform customer problems into novel solutions.

Field engineers in this COE speak Arabic, English, French, Italian, Spanish, German, and even Japanese. Every year, over 330,000 students graduate from Egyptian universities. This talent pool is not only uniquely multilingual, but the Egyptian government is strategically investing in Egyptian outsourcing of information technology services. The confluence of the talent pool and government investment makes Cairo a growing locale for high-tech talent. The COE is strategically positioned to close the gap between the talents of these graduates and the real market needs in the IT sector (the middle ground between academia and industry).

Cairo is a strategic travel hub; Egypt is where Europe meets Africa meets Asia. It is also home to many local academic institutions. Employees at the COE have already formed relationships with nearby universities, including Cairo's Information Technology Institute, Cairo University, Ain Shams University, Benha University, Tanta University, the American University in Cairo, the German University in Cairo, and the Canadian International College. More public universities will be added to this list, as well as Nile University.

As Egyptian field engineers integrate themselves into the global innovation community, they have full access to the innovation resources found around the globe. They experience a level playing field and established channels for submitting their ideas into the corporation at large.

The impact of the Egyptian COE on the corporation has been immediate, and follows the pattern of reverse innovation already highlighted within the pages of this book. Perhaps the greatest example of this impact can be found with the transformation of EMC's traditional telephony system.

The thesis behind reverse innovation is that established products, systems, and services in the United States don't necessarily meet the needs of customers in developing countries, and that these countries need to create new products, systems, and services. The resulting innovations, if successful, can then be exported back into wealthier nations (including the United States).

EMC's traditional customer telephony system falls into this category, and the emergence of the Cairo COE is triggering a local transformation that will spread throughout the globe.

Egyptian Epilogue

EMC Corporation has long owned its own telephony hardware for customer interaction. The hub of the system is in Hopkinton, Massachusetts. As globalization occurred, more hardware was purchased and installed at two strategic locations: Ireland and Australia. The Irish system would handle customer service in Europe, while the Australian system could service the larger Asia-Pacific region.

With the onset of explosive, global demand for IT customer support, the original telephony design reached its limit of scalability. It was simply too expensive and slow for EMC to own and operate its own telephony hardware. When the Egypt COE came online, it would push the solution past its breaking point. The multilingual, multicultural reach of the Cairo office across multiple time zones required a new solution: cloud-based telephony. "Customer support in the cloud" eliminates the need for dedicated phone systems and provides customer support via Internet-based call routing and laptop-based call answering.

The telephony solution deployed at the Egyptian COE provides a level of customer support capabilities that were not possible using the old system:

- Incoming tech support calls can be answered in the native language of the dialer.
- Calls from given time-zones can follow the sun and be dynamically routed to the right person at the right time.
- Escalation for more complex customer problems are passed up the food chain (and tracked) using cloud-based software.
- Escalation support in the cloud can be conducted by either "warm transfer" or call-back.
- Conversations in the cloud can be recorded, monitored, pulled, include multiple parties at multiple locations, and facilitated by a host of other advanced services provided via software.

The Cairo telephony solution is being pushed to other corporations in a classic example of reverse innovation (Shanghai is next, the United States is not far behind). Cairo, with its unique multilingual, high-tech talent pool, will be a critical hub of customer contact in the new corporate world order. The linchpin location will be a primary source of innovation stimulus; the tidal wave of customer needs and requirements can be converted into new ideas by Egyptian employees. Figure 77 depicts the Egyptian COE functioning within the global innovation ecosystem.

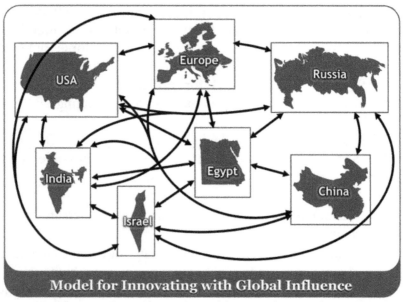

Figure 77. The new Egyptian COE adds great innovation value to the global technology ecosystem.

Multinational corporations have the edge when it comes to innovating with global influence. From Egypt to Russia, from China to Israel, and from Ireland to India, it simply requires a mindset shift on the part of corporate management and corporate employees. The faster it becomes systemic within a corporation, the more quickly that corporation can sell into global markets.

Global employees hold the key. If they find themselves held hostage by the traditional mantra of "innovation only occurs at corporate headquarters," then I offer one last bit of advice.

Your corporation is too big to watch your every move.

Innovate now; apologize later. Your company will thank you for it.

18. Acknowledgments: The Global Thank You

Chronicling the tales of global high-tech intrapreneurs is a story unto itself. I'd like to close this book by describing how it all started and thanking the people who helped out along the way.

My exposure to world innovation started when George Levy of HSM Global invited Stu Miniman and me to the World Innovation Forum in New York City. Thank you George for introducing me to such great thinking!

How does one write a book about the internals of a corporation? It starts with running the idea by a Corporate Branding expert, and Polly Pearson is one of the best in the industry. Any book that broadly covers the global state of internal corporate affairs must reflect well on the corporation as a whole. Polly enthusiastically supported the idea from the beginning and put me in contact with the Volante team.

The Volante team coordinates activities across all of the COEs at EMC, and is run by EMC CIO and Senior Vice President Sanjay Mirchandani. Sanjay graciously extended resources to help with the project, and referred me to Tom Broderick. Tom manages the global framework for Volante and would serve as my main point of contact throughout the project. From east to west, Tom provided the names of folks at each COE who would help me begin to gather stories of local innovation. Eric Silva of Volante also provided me with guidance based on his knowledge of each individual COE.

For China, my initial contact was made with Charles Fan and Peter Madany. Both Charles and Peter are highly collaborative individuals with a strong background in technology and innovation. They provided me with contacts in both the Shanghai and Beijing facilities.

Linda Di, Phoebe Wang, and Li Ying work in EMC's Shanghai facility. I sent them an e-mail asking for an innovation story that involved a complex problem with no easy answer, global collaboration, and persistence to the end. This team came through with a great story and introduced me to Viki Zhang and her United States counterpart, Peter Grogan. Peter filled me in on the critical market need and complexity of the problem that they were trying to solve, while Viki and I exchanged e-mails on her efforts to learn, prototype, and ultimately deliver a great solution with her teammates.

In Beijing, Gus Amegadzie assisted me with introductions to two EMC intrapreneurs: Jidong Chen and Hang Guo. I discovered that Jidong is one of the leading researchers in the Beijing facility, and that Hang survived a rigorous and competitive interview process to land a job at EMC. Within a very short time, Hang not only made a difference within EMC but he has made significant inroads into the industry as well.

The India COE is run by Sarv Saravanan, who put me in touch with Vikas Goyal. Vikas began documenting an extensive set of innovative achievements in Bangalore and was greatly assisted by Niranjan Maka, Amrita Dhindsa, and Kumar Iswaran. There were many stories from which to choose, but the field of information security is so critical nowadays that I decided to further research the Event Source Integrator tool. Arun Narayanaswany spoke with me several times from Bangalore and made strong contributions to the story.

The Israel COE is run by Gil Shapira. Gil enthusiastically referred me to chief scientist and intrapreneur Assaf Natanzon, as well as innovation and M&A lead Gil Goren. Assaf wrote a preliminary draft of the RecoverPoint story and worked closely with Marius Van Handel to edit his first draft. This draft served as the foundation of the story that eventually appeared in this book. Assaf and I had several phone calls so that he could fully educate me on the capabilities, features, and functions found in their innovation. Mark Crooker and Bill Beister provided me with their perspective of Israeli collaboration with the United States and Russia.

It is with a great deal of bias that I write about my co-workers in St. Petersburg, Russia! I have worked with Ivan Gumenyuk for several years and have visited the COE several times myself. The Russia COE is run by Slava Nesterov. Slava referred me to Marina Varzar and Maxas Volodin. They informed me of the effort put forth by intrapreneur Denis Kiryaev on the Captiva project. Denis spent several hours informing (and correcting) each draft of my story. Over the course of writing the book, EMC hired Pavel Egorov, whose advice on the advancement of innovation at COEs had proved invaluable in several chapters of this book.

Finally, I'd like to thank Bob Savage and the team in Cork, Ireland. Cork is another EMC facility that I have had the pleasure to visit. Bob asked Brendan Butler to gather stories and Brendan put a great deal of effort into the search for recent examples of employee innovation (of which there are many in the history of the Cork facility). Eventually, Brendan uncovered the team of engineers who solved the customer problem pertaining to returning failed disk drives. I chose this story due to the increasingly difficult problems presented by government regulations in multiple countries. Shane Cowman proved invaluable in explaining this story to me, and Sean ODonovan provided me with background on the Cork area.

At the eleventh hour, Ron DiPierro educated me on the Egyptian "telephone in the clouds." Its impact on the entire corporation struck me as an important example of global innovation that began as an in-house need and is now being marketed to customers as well. Rasha Olama, my co-worker from the Egypt COE, gave me valuable input on the rise of high-tech in her country.

My Spanish colleague, Vicente Moranta, introduced me to one of his professors at North Carolina State University. My subsequent conversation with Lynda Aiman-Smith strongly influenced my understanding of boundary spanners.

I am extremely grateful to Vijay Govindarajan. Vijay's reverse innovation vision inspired this book, and he also graciously agreed to write the foreword. We spent several hours collaborating over the phone. His enthusiasm for the topic and the book was motivational.

Thanks to everyone mentioned above, and to everyone who worked behind the scenes to make it happen.

When I started my career as a software engineer, the possibility of global collaboration never crossed my mind. Over the course of time, it became quite common to work with a set of teammates in another geographic location. However, I did not and could not imagine that the world would advance to the point where I'd have the privilege to function as part of a global team of intrapreneurs.

In this book, I could only cover a fraction of the innovation that is going on across geographies at EMC. The list of global locations continues to grow, and I fully expect the number of innovations will continue to grow, too.

Building products for the information industry has become a career sweet spot for many of us. There's no greater thrill than diagramming a new idea onto a whiteboard and then guiding it into the hands of customers.

The innovation landscape has forever changed, and I, for one, plan on leveraging the global opportunity.

ABOUT THE AUTHOR

A native New Englander, Steve Todd entered the information storage industry as a student intern while earning his BS (and subsequent MS) in computer science at the University of New Hampshire. After a few years of work at that same company, Steve was recruited – repeatedly! – to interview with the late scion Dick Egan (the "E" of EMC). Eventually he joined the ranks of an organization that came to be known as the fourth horseman of the Internet.

Consistently demonstrating the traits of an innovator – a high-tech intrapreneur – Steve's credits include more than 160 patents filed, the honorary title of Distinguished Engineer, and the reputation of a go-to guy capable of mentoring other would-be innovators.

In this, his second book, Steve elaborates on the makings of an intrapreneur and invites enterprises to look beyond their provincial boundaries to expand their global opportunities, grow new revenue streams, identify and solve customer problems, and nurture ideas into marketable products. The answer, as Steve has repeatedly proven, is to empower intrapreneurs.

Steve's first book, *Innovate With Influence,* is a personal handbook for innovators. He also discusses high-tech innovation on his blog, the *Information Playground.*

Steve and his wife, children, and golden retriever reside in eastern Massachusetts.

www.ingramcontent.com/pod-product-compliance
Lightning Source LLC
Chambersburg PA
CBHW051238050326
40689CB00007B/978